An Introduction to the Philosophy of Science

An Introduction to the Philosophy of Science

LISA BORTOLOTTI

polity

First published in 2008 by Polity Press

Polity Press
65 Bridge Street
Cambridge CB2 1UR, UK

Polity Press
350 Main Street
Malden, MA 02148, USA

ISBN-13: 978-0-7456-3538-5
ISBN-13: 978-0-7456-3539-2(pb)

A catalogue record for this book is available from the British Library.

Typeset in 9.5 on 12pt Utopia
by Servis Filmsetting Ltd, Stockport, Cheshire
Printed and bound in Great Britain by
MPG Books Ltd, Bodmin, Cornwall

For further information on Polity, visit our website:
www.polity.co.uk

To Rita, who has always helped me

Contents

Figures

Tables

Acknowledgements

Writing this book has been more of a collaborative enterprise than it might seem at first glance. In the course of the following chapters I review classic debates in the philosophy of science, but also focus on specific arguments that I have developed with other philosophers and in particular Matteo Mameli (on methodological deception in psychological research), Bert Heinrichs (on delimiting the concept of research), and John Harris (on the ethics of enhancements).

I am also indebted to Ángel Fernández, Asja Portsch, Francis Longworth, Maggie Curnutte and Nigel Leary for many helpful suggestions. Nigel has been an absolutely fantastic help at various stages of the project and he is entirely responsible for enriching the bibliography on natural-kind terms and especially on "jade". I am very grateful for his competence, his valuable assistance and his enthusiasm.

I wouldn't have written this book without the encouragement of Keith Maslin (Esher College), Emma Hutchinson (Polity Press), and my wonderful head of department, Helen Beebee. Writing the book would have been much more difficult without the constant support of my very understanding parents and of my friends. For being there and gently pushing me towards the finish line, I thank Yujin Nagasawa, Matteo Mameli, Matthew Broome, Dan López de Sa, Jordi Fernández, Edoardo Zamuner, and Esa Díaz-León.

I am also grateful to all the people who taught me to love philosophy in general and philosophy of science in particular: Maurizio Pancaldi, Eva Picardi, Maurizio Ferriani, Geoffrey Cantor, Donald Gillies, David Papineau, Bill Newton-Smith, Martin Davies, Kim Sterelny, and John Harris (in the order in which I had the pleasure to meet them).

I was lucky to be part of a very stimulating research environment when I was working on the EURECA Project (on Delimiting the Research Concept and Research Activities) at the Centre for Social Ethics and Policy in Manchester from 2004 to 2005. Since I joined the Philosophy Department at the University of Birmingham I have enjoyed fantastic support all around and have tried draft chapters of this book on very patient undergraduate students. Recently I have also had the opportunity to visit the European School of Molecular Medicine (SEMM) at the Institute of Molecular Oncology Foundation in Milan, where I witnessed the inspiring fruits of the happy marriage of science and philosophy.

Long before I discovered philosophy, I promised to dedicate my first book to my sister. Back then, Rita and I could not imagine that the book would have been an introduction to the philosophy of science, but here it is. I hope she won't be too disappointed.

Introduction: What is Science?

This is a guide to the central philosophical issues raised by the practice of science. It is not just for the philosopher who is curious about science, but also for the scientist who wants to know more about philosophy. And for anybody interested in what gives science a special status in spite of the continuity between scientific research and other human activities.

Each chapter centers on a cluster of problems and intends to equip the reader with basic tools for the appreciation of classic debates in a traditional area of philosophical exploration. In chapter 1, on *demarcation*, some of the attempts philosophers have made to answer the question of what makes science special are reviewed and assessed. In chapter 2, on *reasoning*, some strategies for the acquisition and derivation of scientific knowledge are identified and compared. In chapter 3, on *knowledge*, the structure of scientific theories, their formation, their confirmation and the nature of explanation are examined. In chapter 4, on *language and reality*, the language used in scientific theories comes under scrutiny; especially, the distinction between observational and theoretical terms, and the potential linguistic and conceptual barriers for scientific understanding. Also, the purpose of science is addressed: is science aimed at describing how things really are, or just at giving us the means to predict the phenomena we are interested in? In chapter 5, on *rationality*, the nature of theory change and scientific progress are investigated. In chapter 6, on *ethics*, some examples of the complex relationship between science and society are discussed, and questions are asked about the ethical constraints that should be imposed on scientific research. The capacity of science to provide morally relevant benefits to individuals and societies is also briefly addressed.

The volume emphasizes two areas: (1) the acquisition, systematization, and revision of knowledge in science; and (2) the intricacies of the relationship between science and the rest of society. You will read about classic and current debates on scientific reasoning and rationality in science, and be invited to reflect at all stages on the authority and responsibilities of those who promote science and engage in scientific research in our society.

You might still wonder: "What can I hope to achieve by reading an introduction to the philosophy of science?" Although we are bombarded with information about what scientists do and how science affects all aspects of our lives, we rarely stop to reflect on the significance of scientific research, its status and how it differs from other human activities. In the course of our education and in everyday life we cannot but gain some understanding of science. By watching a documentary on the fossils in the Galapagos islands; by hearing news of a recent breakthrough on avian flu; by reading about black holes in popular science books, we get acquainted with the struggles and results of scientific research and expand our knowledge of nature. But confronted with the variety of methods and objectives of scientific investigation, its successes and failures, we find it extremely difficult to capture what makes the practice of science distinctive.

Trivially, if nobody had ever systematically engaged in the empirical investigation of nature, today we would not benefit from many of the technological advances that characterize our lifestyles, such as vaccination, earthquake prevention measures, and mobile phones. Almost everything around us – clothes, food, buildings – would not be here (at least in their present form) unless people had invested valuable time and scarce resources in the making of science. But scientific research has not only affected the lifestyle of many human beings. The outcomes of scientific research have shaped our beliefs about the world by changing our conception of ourselves and of the differences between human beings and other living beings on earth. Influencing our systems of beliefs, important elements of the so-called scientific method have fed back into the style and form of our everyday thinking. We believe that rationality requires us to predict future events on the basis of our current knowledge. We value explanations for the events we observe if they are comprehensive and consistent with the available evidence. When faced with problems, we find solutions that rely on our past experience and we become better at solving them with time. We do change our minds when our experience does not support our initial beliefs. Even if we rarely or never reflect upon the way in which we form opinions and account for the facts that matter to us, we record information, learn from our mistakes, revise our beliefs, and improve the predictive and explanatory power of our theories. In some weak sense, we all are – or strive to be – everyday scientists.

These observations lead us to a tension in our conception of science. On the one hand, scientific research seems to be unique among human activities and invested of special significance and responsibility. Some would even claim that the pursuit and achievements of science are *the* mark of humanity. In many contemporary societies, science is authoritative and scientists are the experts consulted by governments in time of emergency and for the purposes of forward planning, improving quality of life and preventing natural catastrophes. On the other hand, the objectives and methods of scientific research are so intertwined with the other objectives and methods that it is a serious challenge to single out the features of scientific research that make it truly distinctive.

By thinking systematically about science we might be in a better position to solve the tension between the uniqueness and the pervasiveness of science. In this introduction to the philosophy of science, we shall revisit some classical debates in philosophy about rationality and reasoning, the formation and justification of theories and the nature of reality and progress. We shall also explore current debates about the way in which our concepts carve up nature and ethics and science are mutually constraining. Embarking on this journey can help us come to a more informed and less muddled view of what science is and why it matters.

This is a journey for the absolute beginner, and help is available in the form of questions and exercises to check comprehension and direct further research; tables to facilitate understanding and provide illustrations of some of the points made in the text; examples from the natural, social, and medical sciences; questions to invite reflection, shape group work, stimulate debate, or guide essay writing; some resources for further reading at the end of each chapter; a comprehensive thematic bibliography in the end; and a substantial glossary of technical terms which contains also brief biographical details of important scientists and philosophers who appear along the way.

Enjoy!

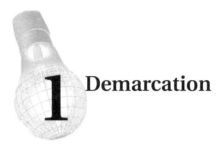

1 Demarcation

There is much skepticism about the possibility of effectively distinguishing science from non-science. The view that we cannot provide a satisfactory demarcation criterion is motivated by the failed attempts to provide such a criterion in the past and by the observation of the ever-increasing diversity of methods and aims of those disciplines that we are inclined to regard as scientific. How can we hope to offer a unified account of what makes research scientific in disciplines as different as physics, geology, and economics?

Even though the enterprise of delimiting science seems to be hopeless, there are very good reasons to keep trying. It is important to know which experts to trust, which research projects to fund, which theories to teach in schools. And decisions about these issues cannot be made only on the basis of theoretical consistency or the apparent adequacy of the theory to empirical data. We need an account of what science is, what scientists do and what aims and methods characterize scientific research. The successful account (if there is one) is not likely to be very specific, as it is true that specialization has led to a range of different concepts of evidence and, moreover, different criteria for success within the natural sciences and between natural and social sciences.

The issues involved in delimiting science acquire great importance in contemporary society where science is invested of special authority and responsibility. Scientists are often the experts who advise politicians on policy making and their opinions are solicited and listened to widely in the media. In virtue of their expertise, of their status as scientists, some of them are called to find solutions for many of our everyday problems, from handling the effects of droughts to preventing newly developed educational curricula from adversely affecting young children. If so much responsibility is placed on individual scientists and on the scientific community at large, we seem to require with some urgency an account of what counts as a *proper* scientific discipline, as opposed to the exercise of disciplines that do not share the same respectability and social authority, such as astrology and chiromancy. Moreover, the pursuit of scientific research in many areas (e.g. research in biomedicine, agriculture, renewable sources of energy) can bring great benefits to individuals and societies, and therefore should be widely supported and promoted. If there is a value in science in the context of limited public resources, it is pressing to be able to identify genuine instances of scientific research and worthwhile research projects.

Traditionally, the discussion about the demarcation criterion between science and non-science has been structured around the attempt to explain why physics is a science and astrology isn't, and in what ways the scientific method differs from magic or divine revelation. But philosophers who are interested in the demarcation criterion today have in mind a cluster of interrelated questions and do not necessarily aspire to provide a description of science that can answer all of them at once.

Here you have a provisional list of questions:

- Does the subject of investigation matter to whether a research project is deemed as scientific?
- Can anthropology, psychology, and economics be regarded as legitimate sciences, even if they are not governed by laws?
- Creationism has the superficial appearance of a science. Why is it not regarded by many as a legitimate scientific enterprise?
- What is the difference between philosophy and science, given that they are both aimed at a better understanding of the phenomena around us?

In the twentieth century philosophers inspired by a movement called Logical Positivism analyzed ways of obtaining and organizing knowledge with a view to identifying important differences between science and metaphysics, and science and ethics. The logical positivists, many of whom had been trained in the natural or social sciences or in mathematics, strongly believed in the value of science (that is why they are called logical *positivists*) and attempted to justify its status as the only respectable source of factual knowledge by analyzing the logical structure and language of knowledge claims (that is why they are called *logical* positivists). One objective of this chapter is to review and assess the strengths and limitations of their account of demarcation between science and non-science before examining later developments of their views and the objections that their account raised.

Some of these objections can be found in the work of Karl Popper, Paul Thagard, and Paul Feyerabend. Popper, who shares some of the logical positivists' emphasis on the value and objectivity of science, has a different focus in his search for a demarcation criterion. He believes that science is the rational enterprise *par excellence* and actively looks for a workable strategy for telling genuine scientific theories apart from theories that might at first glance appear as scientific but fail to be so (instances of *pseudo*science).

In contrast to Popper and the logical positivists, Thomas Kuhn highlights the historical and social factors that determine the success of a scientific theory or a research project. A theory or project might be regarded as scientific in one historical and social context and not in another, because the criteria that a theory or project needs to satisfy in order to count as scientific are also subject to change. On the basis of Kuhn's historically sensitive analysis of science, Thagard develops a context-dependent demarcation criterion that attempts to explain how some disciplines can move from scientific to pseudoscientific status, or vice versa. Feyerabend endorses a more radical position by denying *any* special status to science. He argues against the presumed supremacy of scientific methodology over alternative traditions of thinking.

After this brief selective history of the demarcation criterion, I shall draw some conclusions about the recent developments of the debate and put forward a suggestion for delimiting research activities.

By the end of this chapter you will be able to:

- Point at some differences between metaphysics and science, and between ethics and science in the light of the considerations put forward by logical positivists.
- Explain and assess Popper's attempt to provide a demarcation criterion between science and pseudoscience.
- Appreciate the social factors that might contribute to the change in status of a theory and discuss the merits and limitations of an anarchic methodology.
- Discuss and rate different attempts to demarcate science with respect to examples of pseudoscience and bad science.
- Identify the challenges facing the project of delimiting research activities.

1.1.   Science and Non-Science

It is a matter of controversy when humans began to do science, as we shall see in the next chapter when we trace the origins of the so-called experimental method. In all civilizations there have been people interested in describing and explaining natural events, such as the motion of celestial bodies, childbirth, or the occurrence of floods. They made hypotheses and reached conclusions after completing a series of observations about the phenomena they wanted to explain. In a broad sense, they were doing science. But one accepted view is that *modern* science has a special character that cannot be found in the previous attempts to explain natural phenomena. What this special character consists in is a matter of debate, but one widespread assumption is that an evidential basis is necessary in order to consider a hypothesis as scientific.

Since Aristotle's *Physics* (350 BC), scientific reasoning has consisted in forming hypotheses to explain some observed event and in revising explanatory hypotheses if future observations fail to support them. Whether this process counts as providing an evidential basis, is also open to interpretation. Did modern science begin when humans went beyond the *passive* observation of nature and started to *actively* intervene in natural phenomena? Today the manipulation of nature is common in many of the sciences where experimenters create special conditions for an event to occur in order to control for variables and refine their hypotheses. But for a long time in the history of human investigations of nature, the basis for the accepted theories was constituted mainly by thought experiments and unaided observations, and the contemporary distinction between science and philosophy was at best blurred.

One might argue that reliance on evidence and even active manipulation of nature are not sufficient criteria to tell modern science apart from other evidence-based theorizing. A series of hypotheses that explain the occurrence of the phenomena we are interested in does not constitute scientific knowledge unless these hypotheses *hang together* in the right way. Tested hypotheses need to be part of a structured and coherent system in order to contribute to the body of scientific knowledge.

The formulation of explanatory hypotheses, the manipulation of nature in order to refine and test them and the formation of coherent theories are some of the things scientists do. But is there a list of necessary and sufficient conditions for a body of knowledge to be genuinely scientific, or for an activity to count as scientific research?

In this section we are going to consider in some detail the reasons why scientists and philosophers are thought to be engaging in different enterprises.

Exercise: Before reading on, write down three ways in which science and philosophy differ, given your understanding of their respective methods and objectives.

1.1.1.   *Analytic and synthetic statements*

Statements are of different kinds. Some statements are *synthetic*, that is, we would not be able to tell whether they are true or false just by reflecting on their logical structure or on the meaning of the terms contained in them. "Today it will snow" is a synthetic statement. Other statements are *analytic*, that is, they are either true or false in virtue of their logical structure or the meaning of the terms contained in them. "A square has four equal sides" is analytic, as squares have by definition four equal sides. "Today it will either snow or not snow" is analytic, because it is a disjunction of mutually exhaustive statements.

Logical positivists thought that all synthetic knowledge must be acquired and verified through experience (that is why they are also called logical *empiricists*), whereas experience is irrelevant to the acquisition or verification of analytic knowledge. But not all the examples of synthetic statements seem to work this way. There are some synthetic statements, those we call normative, such as "killing is wrong," which are not always true or false by definition, but whose truth or falsehood cannot be easily established via empirical investigation. Other synthetic statements do depend for their truth or falsity on how things turn out to be, but we cannot think of a way of putting them to the test, of verifying them. "Being a philosopher was an essential property of Aristotle" – this statement does not satisfy the conditions for analyticity, but it is hard to say what *empirical* tests could determine its truth or falsehood. One can argue for its truth or falsehood on the basis of some theory about what counts as an essential property, but not on empirical grounds (although there are more and less useful notions of what an essential property is given how the world has turned out to be).

The traditional view is that there is a fundamental difference between the descriptive and the empirical on one side, and the prescriptive and normative on the other. The natural sciences are about *facts*. How hot is the water when it boils? How fast is the acceleration of a falling body? How old is that fossil? Why do earthquakes happen? What causes a chemical reaction? Why do primates use alarm calls? However, in other disciplines, we also describe and explain *facts*. Which are the most common metaphors for death, and are they shared by different cultures? What were the effects of World War I in Europe? Why did artists start using perspective in the 15th century?

Although many disciplines seem to study facts, describing and explaining how things are is not all we do. Sometimes we want to know how things *ought to* be on the basis of some principle or norm. Is democracy the best form of government? Is killing intrinsically wrong? Should downloading music from the internet be a criminal offence? Forms of government and instances of human behavior are objects of evaluation and can be good or bad, adequate or inadequate, right or wrong. These normative claims arguably cannot be justified on the basis of mere experience.

For the logical positivists what distinguishes scientific statements from statements of logic, philosophy, religion, literature, etc., is that they are synthetic *and* that their truth can be established via empirical tests (i.e. they are verifiable). In their view, the possibility of conceiving of evidence that can confirm or disconfirm a given statement is what makes a synthetic statement *meaningful*. And synthetic statements that cannot be confirmed or disconfirmed via empirical evidence have no meaning at all. Experience can confirm the claim that the water in the kettle is boiling. But what straight-forward observation can confirm the claim that killing is wrong? Logical positivists are not happy with the traditional thought that some statements are neither analytic nor verifiable via experience and want to find a way of accounting for the apparently puzzling nature of these statements.

In what follows we shall review some of the implications of the view that only synthetic statements that can be verified are meaningful, and reflect upon the way in which logical positivists characterized the difference between science and ethics, and between science and metaphysics.

1.1.2.   The "elimination" of ethics

For Alfred Ayer (1936), who shared and divulged many of the ideas that the logical positivists put forward about the distinction between science and philosophy, ethical statements cannot be verified or falsified by appealing to experience. According to him, this explains why ethical issues generate endless disputes that are ultimately unfruitful. When we think about ethical statements, says Ayer, we have the impression that we need to latch onto how things should be, to their normative dimension. But that appearance of normativity in ethical statements, according to Ayer, is just an illusion. There is no normative dimension to ethical statements; there are only preferences that are ultimately subjective and often clash with the preferences of others.

Ayer argues that ethics as a normative discipline has no reason to be. What is at issue in ethical debates is the expression of preferences that are partly determined by psychological and cultural facts about the individuals or groups who express them. When I claim that killing is wrong, all I am saying is that killing is not a practice I approve of, because I have a negative emotion associated with it ("Killing is wrong" means just "Boo to killing!," Ayer would say). And this negative association is partly determined by the fact that I have been brought up in a social context in which unnecessary killing is largely condemned. Ayer's conclusion is that ethics should not be seen as an independent subject issuing normative claims, but should be rather subsumed under empirical sciences such as psychology or sociology.

Ayer's views on ethics are radical and controversial. To consider some alternatives to his position, we would have to explore the vast philosophical debate about the nature of ethical facts. But for our purposes here what is relevant is that in his view: (1) scientific statements are synthetic statements that can be verified; (2) ethical statements can be viewed either as synthetic statements that cannot be verified (and therefore are empty of meaning and a waste of time) or as synthetic statements about individual or societal preferences that can be studied empirically by the psychological or social sciences.

 Exercise: Of the following sentences, which represent reality and which express feelings or preferences?

- *Anxiety causes depression.*
- *All events have a cause.*
- *Playing cards is a waste of time.*
- *Carrot juice is good for you because it contains vitamin C.*
- *Going to war was a mistake.*

Hans Reichenbach (1951), another logical positivist, arrives independently at a very similar conclusion to Ayer, and he also does so by reflecting on the nature of what seem to be ethical statements. He argues that these linguistic expressions are not genuine statements, because they do not describe how things are but rather issue *directives* or manifest *wishes* and therefore cannot be true or false. Saying that killing is wrong is either equivalent to the imperative "Do not kill!" and it is a linguistic utterance that people use to influence or control other people's behavior; or it is the expression of a preference for a world where there is no killing. Whereas statements that can be verified have empirical or cognitive significance, directives or wishes have only *instrumental* value, as they are a way for the speaker to achieve something she wants or express a preference. Ethics is not the "science of the ultimate good": it does not contribute to scientific or empirical knowledge at all and it is not *about* the ultimate good, whatever that may be. It is an expression of the will of an individual or a group to influence the conduct of others.

The logical positivists defend very radical views about the status of ethics because they tend to see all acquisition of genuine knowledge as a fundamentally empirical enterprise and they impose verifiability as a condition for meaningfulness on all statements that do not qualify as analytic. The normativity of ethical statements is interpreted by them as an illusion created by the way in which language is (often improperly) used. In their view, the analysis of linguistic utterances is a means to disclose the alleged nature of ethical statements and provides a demarcation between them and legitimate scientific statements.

1.1.3.   Metaphysics as poetry

How should we conceive the distinction between science and metaphysics? There is a sense in which both the natural sciences and metaphysics are aimed at gaining a better understanding of nature. It is interesting to reflect on the history of the relationship between science and metaphysics, because thinkers who have immensely contributed to the progress of science, such as Isaac Newton or Albert Einstein, did express metaphysical views and worked under explicit metaphysical assumptions.

For the logical positivists, the difference between science and metaphysics lies in the methods by which the investigation of nature is conducted and in the significance of the claims that are formulated within these disciplines. Take for instance the Greek philosopher Plato. In many of the dialogues he wrote, he claims that the world of our experience, including the chairs we sit on and the sun we see rising and setting everyday, is only half real. Ultimate reality is made not of material objects but of *forms*, or ideas, that we cannot see or touch, because they inhabit a different world from the world of our experience and cannot be perceived through our senses. But, if the forms cannot be seen or touched, then we cannot know on the basis of our senses whether they exist and whether they have the attributes that Plato ascribes to them.

The logical positivists believed that metaphysical claims such as "The world of Forms cannot be perceived through our senses" do not have any empirical or factual meaning, because they are not analytic and they receive no evidential support from experience. Their view is that most metaphysical claims have no meaning whatsoever and are misleading, as these statements deploy words that commonly refer to the objects we can experience with our senses in order to describe objects that are often by definition outside or beyond that experience.

Rudolf Carnap (1935) compares a statement about the existence of Platonic forms with a statement about the existence of kangaroos. He notices that when zoologists assert that kangaroos exist, their assertion can be verified because it follows from it that at certain times and places things of a certain sort can be observed. Plato's assertion that forms exist is different, as forms can *never* be perceived. Carnap believes that metaphysical claims such as "Forms exist in a spaceless and timeless sphere" do not represent reality and thus they cannot be either true or false. Rather, they express something, such as the desire to believe in entities that are not as subject to change and destruction as physical objects are. The desire expressed by a metaphysical claim has no scientific or theoretical content and can be compared to the work of a poet. But there is a difference between the attitude of the metaphysician and that of the poet. The poet knows she is describing feelings and desires in her writings, whereas the metaphysician is deluded and she mistakenly believes herself to be contributing to a form of factual knowledge. Evidence of this delusion is that the metaphysician is prepared to enter into an argument with other metaphysicians about the truth of claims about objects or properties which cannot be experienced. For Carnap, metaphysical claims are expressive, not representational, and only *appear* to have theoretical content to those who endorse them.

 Discuss: Is Carnap's distinction between expressions of feelings and desires and representational statements convincing? Is it useful?

Karl Popper (1959, 2002) disagrees with the logical positivist idea that metaphysical claims have no representational meaning. He notices how some metaphysical hypotheses have had an important influence on the development of scientific hypotheses. His example is atomism. The theory that all matter is composed of indivisible parts ("atoms") emerged in Ancient Greece, first formulated by Leucippus (c. 500 BC) and Democritus (460–370 BC). This theory was the outcome of philosophical speculation, and developed as an attempt to solve paradoxes about motion and change. It remained a metaphysical hypothesis about the nature of reality for a long time: different versions of it were articulated in the seventeenth century by philosophers who were interested in the ultimate nature and composition of matter.

One could argue that from the nineteenth century atomism *became* a scientific hypothesis, developed by John Dalton in organic chemistry and James Maxwell in relation to the kinetic theory of gases. In the twentieth century the existence of atoms stopped being controversial. Of course, the atoms whose existence we now accept are described very differently from the atoms Leucippus and Democritus first talked about, but someone might argue that atomism as a scientific hypothesis would not have emerged in the absence of the previous metaphysical tradition. Popper considers this as a difficult case for whoever insists that metaphysical hypotheses have no representational meaning. He argues that even from myths one can derive hypotheses that are subject to empirical testing: the Copernican

Table 1.1 Development of the conception of the atom.

Greeks (500 BC)	Atoms introduced as the indivisible parts of matter.
Thomson (1904)	Discovery of the electron as a subatomic particle.
Rutherford (1911)	Proposed model of the atom in which the nucleus has positive charge and is surrounded by orbiting electrons with negative charge.
Bohr (1913)	The energy of the particles in the atom is quantized (electrons can occupy only certain orbits).
De Broglie, Heisenberg, Schrodinger (1924–6)	Electrons act both as particles and waves and we can never determine both their positions and momentum in the atom.

system, for instance, was inspired by the Neo-Platonic fascination with the light emitted by the Sun.

The perceived nature of a metaphysical claim is partially explained by the way they are justified. Metaphysicians of Ancient Greece such as Democritus and Plato did not conduct experiments nor based their ideas on a series of careful observations but achieved their conclusion by reason alone, providing arguments for their views that did not typically have empirical statements as their premises. Contemporary metaphysicians are less inclined to speculate on a world of unobservable objects and properties and rather prefer to make sense of reality in a way that is compatible with, and sometimes even a conceptual aid to, the currently accepted physical theories. One example of this interplay between metaphysics and physics is the study of the nature of time, which has been informed and inspired by the theory of relativity and its significant consequences for the commonsense view of reality.

Even though in metaphysics we do not expect researchers to set up experiments and find empirical confirmation for all the statements in their theories, it is true that some metaphysicians would take into account what physics has unveiled about the structure of reality, elucidate the concepts involved in the explanation provided by the accepted scientific theories and deepen our understanding of such concepts

(Ladyman et al. 2007). That said, the debate about the role of metaphysics is still extremely lively, and philosophical traditions differ with respect to the way in which the relationship between science and metaphysics is conceived.

 Exercise: Can you think of hypotheses that are considered as scientific but have no empirical bases? Should these hypotheses count as scientific?

1.2. Science and Pseudoscience

The logical positivists provided the criterion of verifiability as a criterion for meaningfulness of statements: a statement has meaning if it is either synthetic and can be verified through experience or it is analytic. Genuine scientific statements (e.g. "Heavy smoking increases the chances of contracting lung cancer") seem to satisfy the criterion, as they are synthetic statements that can be verified, but many ethical and metaphysical claims appear as synthetic and yet cannot be verified through experience, so they fail the meaningfulness-test.

Things are more complicated than the neat distinction put forward by the logical positivists might suggest. According to Schlick in his lectures entitled "Form and Content" (1938), Descartes' claim that "Only human beings are endowed with consciousness" cannot be empirically verified. But whether we want to regard claims about consciousness as metaphysical or otherwise depends on what type of justification we can provide for endorsing them. If we have a definition of consciousness that makes it impossible for beings other than humans to be conscious, then the claim is an analytic statement. But if the definition of consciousness does not rule out *a priori* that non-humans can be conscious, Descartes' claim counts as synthetic and we can easily imagine scientifically respectable ways of providing a justification for it. Suppose we thought that certain regions of the human brain were involved in any experience that we deem as conscious and we also came to believe that these regions were significantly different in the brains of non-human animals, or totally absent from them. In these circumstances, we would have some empirical bases to assess the truth of claims about consciousness in non-humans. Descartes' claim would qualify as a verifiable synthetic statement.

Schlick thought this was a good example of an unverifiable metaphysical claim because he assumed that the philosopher who put it forward, Descartes, did not justify it on the basis of empirical data that he could have verified (although Descartes was a keen vivisectionist and had a lot of hands-on knowledge of animal physiology). But the example shows that the distinction between what can be verified and what cannot be verified is not set once and for all, and that seemingly intractable problems can become more open to empirical investigation thanks to the advances of science and technology.

Other challenges have been made to the criterion of verifiability as a criterion for meaningfulness, and also as a criterion of demarcation. There are doubts that the criterion can be sufficient to distinguish between statements which belong to genuine scientific theories and statements which don't. For instance, the criterion does not seem to have the resources to discriminate between synthetic statements that are part of a respectable physical theory and those that appear in a weekly horoscope. Most of the claims made by astrologists are arguably synthetic and some are even subject to verification. These statements satisfy the meaningfulness criterion, and yet we are resistant to accepting them as scientific and often find them lacking in justification and empirical support. So, we need to look further for a way of marking the perceived gulf between physics and astrology.

 Exercise: Before reading on, make some notes about the main differences between physics and astrology.

1.2.1. Is astrology falsifiable?

A major contribution to the classical problem of demarcation was made by Popper (1959, 2002), who believed that science differs from pseudo-science in that it aims at the production of *falsifiable* hypotheses. Popper is not convinced that, in the context of demarcation, appealing to the possibility of verification is satisfactory. His suggestion for an alternative strategy is based on the observation that general statements can never be verified by experience, because for their verification an infinite number of observations would be necessary. How many observations of white swans are needed to verify the statement "*All* swans are white"? General statements of the form "All Xs are Y" concern past, present, and future instances of X and therefore no number of observations of X could ever be sufficient evidence to establish with certainty the truth of that general statement. Of course, if I observe one hundred swans and they are all white, I will reasonably expect the next swan I observe also to be white. But, as we know, on a journey to Australia the observation of a black swan might prove a revelation. The existence of just one instance of X which is not Y proves that the general statement is false after all.

The starting point for introducing the notion of falsification is that a single experience can contradict the prediction based on a general hypothesis and that this is sufficient to prove that the hypothesis is false. According to Popper, only scientific theories are falsifiable this way, whereas pseudoscientific theories and metaphysical theories are immune from empirical failure. Hence, he thought that an appeal to falsifiability was the most promising way to distinguish between science and non-science. Can this insight account for the pseudoscientific status of astrology?

Popper (1963) argues that there is an important difference between (a) *predicting* observational evidence on the basis of a given theory and

(b) *accommodating* evidence so that it turns out to be compatible with the theory. The former practice characterizes healthy scientific enterprises, whereas the latter is typical of pseudosciences. According to Popper, a good scientific theory is incompatible with the occurrence of certain events, and therefore *prevents* certain things *from* happening. In this sense, science is *risky*. Popper illustrates this point with the example of Einstein's relativity theory. The predictions that the theory allows us to make are subject to confirmation and disconfirmation, and if they are disconfirmed, it is bad news for the theory.

Here is another example of a risky prediction. Suppose you are considering a model of fluctuations of the stock market according to which every time there is political instability in a country, share prices drop in that country. On the basis of this model, you predict that the next time there is political instability in Italy, share prices at the Milan Stock Exchange will drop. If your prediction does not come true, the model has been falsified.

As opposed to a scientific theory that makes risky predictions, pseudoscientific theories are virtually irrefutable. No evidence would speak against these theories and lead us to reject them, because they are formulated in ambiguous ways or can be twisted to accommodate any apparently conflicting piece of evidence. One of Popper's favorite examples is psychoanalysis. Any clinical observation can be interpreted in the light of the theory and no instance of human behavior would definitely contradict the hypotheses made on the basis of the theory. Astrology also fits this description: its predictions are often phrased in such general terms that no future event will definitely be in conflict with them, guaranteeing immunity to the theory.

Suppose that you are still interested in predicting the behavior of the stock market. This time you use a different model which tells you that every time there is political stability in a country, share prices change – but it does not tell you whether they go up or they drop. This model is still risky (as it would be falsified if the share prices remained exactly the same during a period of political instability), but it is *less* risky than the model we considered previously, because it does not specify in what ways prices would change, and therefore it is immune from some cases of empirical disconfirmation. The model would be pseudoscientific for Popper if there were no circumstances in which it could make predictions that turned out to be false. To sum up, pseudosciences for Popper are not genuinely open to falsification because no event is obviously ruled out by them.

Exercise: Do some research on two of the following and then decide if they meet Popper's criteria for pseudoscience – homeopathy, phrenology, archeology, ufology, evolutionary psychology.

Critics of Popper have undermined falsifiability as a demarcation criterion between science and pseudoscience on the basis that some elements of a scientific theory (such as laws in theoretical physics) are not directly falsifiable, whereas a pseudoscience such as astrology can generate falsifiable statements. If these critics are right, then falsifiability is neither sufficient nor necessary for demarcation.

It is not sufficient because there seem to be falsifiable hypotheses that are not scientific. For instance, Paul Thagard (1978) reports that there have been some attempts to provide empirical confirmation via statistical methods of the view that the position of the planets at the moment of an individual's birth is correlated with the choice of occupation of that individual later in life. To discover that a person's birth is not correlated to her occupation in the way that astrological theories indicate could in principle count as a falsification of the theory.

Falsifiability is not even a necessary criterion of demarcation. Alan Chalmers (1999) reminds us that predictive failure does not always indicate that a scientific theory is proven wrong. As we shall see when we discuss scientific theories in chapters 3 and 5, even if observations seem to contradict the tenets of a theory, in the practice of science it is sometimes perfectly acceptable to preserve the theory and modify instead the auxiliary hypotheses that we need to combine with the theory in order to make it testable (Lakatos 1970; Kuhn 1962, 1970; Kuhn 1996). There can be scientific hypotheses that are so strenuously defended by the scientists testing them that they are made to resist attempts of falsification in the face of inaccurate predictions.

1.2.2. Context-dependent factors in demarcation

Inspired by Kuhn's historical and social analysis of science, Thagard agrees with Popper that astrology is a pseudoscience, but argues that the reasons why astrology is a pseudoscience are not exhausted by the application of the falsifiability criterion. In order to determine the status of a discipline, we also need to examine some characteristics of the community of practitioners pursuing that discipline and the historical context in which these investigations are conducted. A healthy scientific discipline has a community of practitioners largely in agreement about the main principles and methods that characterize that discipline. The practitioners are seriously concerned about apparent disconfirming evidence, attempt to find solutions to the misfit between theory and data, and actively engage in the severe testing of the theory. Both the stage of development of the discipline and the acknowledgement of the competition do matter to its status as a science. Has the dominant theory been struggling for a long time with apparent counter-evidence? Are there other theories available that can explain the relevant phenomena in a more satisfactory way?

According to Thagard, the reason why astrology is in bad shape today is that practitioners of astrology have failed to make significant progress in some time and that we have now more successful and reliable ways to explain human behavior within cognitive and social psychology. Thagard does not rule out that, at some point in the past, e.g. before the development of psychology, astrology could have been regarded as providing a genuine scientific explanation and prediction of human behavior. Today, though, practitioners of astrology do not make any attempt to develop solutions to the problems the discipline faces, do not actively engage in the severe testing of their theories, appear to be selective in the way in which they consider the evidence in support of or against their claims, and do not compare their explanatory framework with alternative ones. According to Thagard, these symptoms suggest that today astrology fails to qualify as a science.

Discuss: Do you agree that the historical context matters to whether a discipline is regarded as genuinely scientific? Consider, for instance, chemistry and psychology.

1.2.3. "Anything goes"

In the 1975 September/October issue of the *Humanist* a statement about astrology underwritten by 186 scientists and learned individuals appeared. In this statement they argued that modern concepts of astronomy and physics and the science of psychology do not lend any support to the view that the position of the planets can in any way affect the lives and behavior of human beings.

Paul Feyerabend (1979) claims that the statement does not contain any convincing argument to support the view that astrology is less respectable than any of the other disciplines mentioned. Feyerabend concedes that most of the contemporary practice of astrology is aimed at "impressing the ignorant" and does not constitute an example of progressive research, but he challenges the way in which the leading scientists involved in the statement attempt to ridicule it. In the statement, it is argued that astrology arose from magic and that its original tenets do not receive any confirmation in contemporary science. Feyerabend replies that, if this is an objection, it is an objection to the scientific status not only of astrology, but of many other disciplines usually regarded as paradigmatic instances of science. Alchemy, which was not devoid of magical references, is the precursor of modern chemistry.

Feyerabend (1975) defends a view according to which science is just a tradition of thought among many others and it is not characterized by any distinctive and rigid methodological rules. The historical development of science has shown that a variety of approaches have been taken to matters that we today call scientific and that exactly this variety of methods has made progress possible. Referring to some examples of scientific practice in different disciplines and at different times, Feyerabend attempts to show that we are deluded if we think that there is a single method that unifies all scientific enterprises. Rather, he claims that the Laws of Reason that we commonly take to be part of the scientific method, including the idea that scientific theories have a tight connection with reality through observation and experiments, are just a *post-hoc* rational reconstruction of scientific methodology and that they are divulged for the purposes of political propaganda.

In our societies, Feyerabend argues, scientists have a power that is bestowed on them on the basis of the fact that they are depositary of a rational method for the investigation of reality. In order to maintain their power, they provide a distorted image of their way of thinking as superior and dismiss alternative traditions of thought. In chapter 5 we shall review and assess arguments for and against the rationality of scientific progress and we shall discuss these issues further. Especially, we shall think about whether there can be objective criteria for ranking different methodologies, and whether contemporary science really does provide us with a style of thinking that is superior to that of other traditions of thought. For our present discussion, it will suffice to say that thinkers such as Feyerabend believe that no coherent and satisfactory demarcation criterion between science and non-science can be found.

 Exercise: List three reasons in favor of and three reasons against Feyerabend's denial of the methodological supremacy of science.

1.3.   Natural and Social Sciences

The question of the status of the social sciences, whether they are genuine instances of science, seems to revolve around the comparison between their methodology given the nature of the phenomena they are studying, and the methodology of physics. Can we really find sufficient elements of continuity between economics and physics to regard them both as sciences? Popper (1957) distinguishes two approaches to the distinction between natural and social sciences: naturalism and anti-naturalism.

According to the anti-naturalist perspective, there is a gulf between the methodologies of physics and sociology. Here is a very partial list of some of the factors that would suggest a deep disanalogy:

- *Generalizations.* In the physical sciences we generalize from particular facts to universal truths based on the assumption that there are some regularities in nature. But in sociology this procedure does not bear fruits, as circumstances are peculiar to one historical moment in time, and ignoring this aspect would be to ignore the fact that society constantly evolves.
- *Experiments.* In physics experiments represent a way of isolating a phenomenon in order to control some variables and focus on a limited number of relevant factors. In sociology this method would not work, as there is no principled way of deciding which factors are relevant to the research questions to be answered. Moreover, exper-

iments in physics can be repeated in different laboratories and the same results can be obtained, but observations in sociology are always unique as they depend on the characteristics of the fact observed.

- *Complexity.* Social facts are complex not just because variables cannot be easily controlled in artificial situations, due to their historical contingency, but also because the mental lives of individuals matter to the development of society, and in order to understand the explanatory role of these mental lives, psychological and biological facts must be invoked.

- *Prediction.* The claim is that, although making predictions in sociology is possible, it is extremely difficult, due to the complexity of social facts, but also to the effect that making a certain prediction can have on the fact to be predicted. For instance, predicting that a bank will face a financial crisis has an effect on consumers who have trusted that bank with their savings. They are likely to withdraw their money for fear of losing it, thereby compromising further the financial situation of that institution.

- *Objectivity.* The whole relation between the person observing that fact and the observed fact is an issue that concerns the natural sciences as well, to some extent, but seems more pressing in the case of the social sciences. The subject attempting to provide an explanation of a social fact is not placed outside of that fact, in a position of neutrality, but is often embedded in it. An extreme consequence of this claim is that, different from physics, in sociology the scientist's aim is not to disclose truths but to bring about a new phase of social development.

- *Holism.* Following from what anti-naturalists say about complexity and the inadequacy of experiments in the social sciences, there is a further issue which affects the scope of prediction and explanation, that of holism. The thought is that an aggregate in physics might be just the sum of its parts, but a social group is always more than the sum of its members, because personal relationships between members can easily change the dynamics and behavior of the group. The group itself will have a history of its own that is not exhausted in the personal history of its members. That means that when we are attempting to provide an explanation or make a prediction in the social sciences, we always need to pay attention to how particular events or interactions that seem to have very limited and confined consequences determine changes in the whole structure of the social phenomenon to be studied; and we cannot offer localized explanations or predictions, but always need to analyze the totality of the relevant social facts.

- *Understanding.* How do we proceed when we want to understand some facts? If these are natural facts, we probably search for what caused them. If they are social facts, the anti-naturalist says, we search for meaning and purpose. Whereas the former aim, *causal explanation*, can be advanced by observation of regularities and generalizations, the latter aim, *understanding*, requires imagination and empathy.

 Exercise: Can you think of other potential methodological differences between the natural and the social sciences?

1.3.1. *Laws and experiments in the social sciences*

Popper strongly disagrees with the anti-naturalist stance and argues for greater continuity between the methodologies of natural and social sciences. He convincingly argues that the anti-naturalist comparison between physics and economics, or physics and sociology, is based on a naïve and overly simplified positivistic picture of how the scientific community engages in the study of nature.

Although it may be true that generalizations in sociology take a different form from those in physics, it is also true that both can be interpreted as laws or hypotheses that issue a prohibition. For instance, here are two of Popper's own examples: "You cannot build a perpetual motion machine" or "You cannot have full employment without inflation."

 Exercise: Can you think of other examples of prohibitions issued by generalizations in the social sciences?

Popper also claims that the emphasis on holism is misconceived, and so is the rejection of the methodology of those experiments which are aimed at finding out regularities in some aspects of social development, rather than in society as a whole. There are, he says, successful examples of *piecemeal* experiments relevant to the articulation of sociological theories.

Think about the famous experiment on obedience to authority carried out by Stanley Milgram in 1974 (to which we will come back in chapter 6). In an experimental setting he showed that people are strongly inclined to obey figures of authority who tell them what to do, even when the request involves acting in what is perceived to be a morally objectionable way. Milgram wanted to gain a better understanding of what had happened in Nazi Germany where there had been a relatively modest public outrage at the extent to which Jewish individuals and communities were persecuted. His hypothesis is of great generality, as it can be applied to different people in different societies in different historical contexts: people find it difficult to disobey orders imparted by authority figures. And yet his experiment was conducted in a lab, with the research methodology of the social psychology of his time. The experimental results confirmed his hypothesis and generated a lively debate on the constants of human behavior thereby contributing to a better understanding of the dynamic of obedience and resistance in authoritarian regimes.

Exercise: Can you think of other experiments that have been valuable to social sciences?

Experiments such as Milgram's can contribute to the acquisition of generalizable knowledge. The method that is used, Popper says, is the method that he advocates for all sciences: trial and error. We attempt to solve a problem given a certain hypothesis and we might fail or succeed, but what really matters is that we learn from the mistakes we make. If the hypothesis does not seem to work, it is revised or rejected and new tests are made. The difficulty with approaching experiments holistically is that if we test hypotheses that concern the whole of society and get it wrong, it is extremely hard to know exactly what the mistake was. Instead, isolating variables, when at all possible, seems to be useful both in the physical and the social sciences.

There are also other elements of continuity with respect to the issue of experimentation. In physics as well as in other sciences, there are potentially very revealing experiments that cannot be conducted because of methodological or technological limitations, and in those cases often scientists have to do the experiments in their head and use their imagination to predict what the results could be, rather than perform actual experiments (as we shall see in the next chapter). Not even with respect to the common use of thought experiments does there seem to be a gulf between natural and social sciences.

Harold Kincaid (2004) defends a naturalistic view, and argues that there can be laws in the social sciences. But his perspective is different from that of Popper. Instead of identifying laws with statements issuing a prohibition, he describes them as statements identifying relevant causal factors. The complexity of social phenomena does not seem to be an obstacle to identifying causal factors that contribute to an explanation of social facts. For Kincaid, there is no good reason to think that the notion of understanding in the social sciences has to be conceived as dramatically different from the notion of causal explanation in the physical sciences.

The anti-naturalist view is that in the social sciences the "objects" investigated are people with free will and ways of conceptualizing the world, and they are not inert matter. This would determine the type of explanation sought for the behavior studied. Human behavior, the view goes, cannot be explained by the same principles as the behavior of physical objects and requires an effort of interpretation that takes into account the perspectives of the people whose behavior is studied (Taylor 1971). Kincaid does not want to rule out that some social facts (e.g. a ritual) are better accounted for by reference to their significance, rather than to what brought them about, but this does not mean that the search for causes is doomed to failure or irrelevant to the purposes of explanation in the social sciences. After all, what the social sciences aim to study is not just the behavior of *some* individuals at *some* time, but the nature of institutions and the development of large-scale (often recurrent) phenomena. Empathetic understanding might sometimes be required to view a certain situation in the same way as the people embedded in that situation, but this interpretive activity does not preclude other methods to ascertain a subject's perspective, which are based on human psychology, for instance, and which can lead to conclusions that are generalizable to some extent.

1.4.   What is Scientific Research?

In this section the question of demarcation will be approached from a different angle. Instead of searching for an account of science as a unified and static body of knowledge or an account of what makes a discipline scientific, let's consider another demarcation project. What characterizes a human *activity* as an instance of scientific *research*? Three distinct sets of questions seem to emerge when we are considering potential answers to this question. First, an activity that counts as research has a *methodological dimension* and is systematic rather than random. Second, an activity that counts as research has a specific *function* and is aimed at contributing to a body of knowledge. Third, activities that count as instances of scientific research have some *sociological aspects* in common, such as the role scientists play in settling disputes about empirical issues and the way in which new generations are educated in the sciences.

Exercise: The sociological dimension of research will not be explored here, but you can reflect on and discuss the following issues. (1) Can anyone do research, or is some form of training or status required? (2) Does it matter where the investigation is conducted, how it is funded and whether it fits a wider project recognized by a community of researchers?

1.4.1.   Procedural questions

Various procedural issues are relevant to the demarcation of research activities. Research activities tend to be systematic and follow some method whose prescriptions by and large will depend on the discipline within which research is conducted. Whereas natural and social sciences might require rigorous empirical testing, other disciplines might require from their standard practices that they are just transparent and open to rational criticism.

When we think about traditional procedural questions, we seem to encounter a tension that is reflected in the development of twentieth-century philosophy of science. On the one hand, science is so compartmentalized and scientific procedures can vary so widely that maybe only specialized scientific communities can determine if some particular activity conforms to the often abstract requirements of the currently accepted methodology. For instance, Max Black (1954, ch.1) observes that when we talk about scientific methodology in general we tend to abstract from what we know about physics, but in astronomy there are no experiments and geography is mostly descriptive. This suggests that there is no hope of finding a very detailed description of the scientific method that fits all the sciences. On the other hand, some demarcation criterion is needed for the purposes of public understanding and policy making. Although it is not realistic to aspire to describe one ultimate method for all the disciplines that can be regarded as scientific, there seem to be essential procedural elements which can help us distinguish research from other activities. Talking about *one* scientific methodology seems misleading, not just for the diversification of scientific disciplines, but also because the method used in scientific research, as well as the scientific theories arrived at by that method, can be subject to revision.

In very broad terms, two methodological requirements seem to apply to any activity that we would be happy to regard as scientific research. First, scientific research must be conducted in a way that allows "reality-checks," that is, testing must be part of the way in which conclusions are reached and justified. Second, both the conclusions reached and the reasoning steps necessary to reach them must be transparent and open to criticism.

These two points seem to match some of the requirements suggested by Popper and Thagard in relation to the perceived differences between the practice of physics and astrology. But notice that, whereas the requirements of sensitivity to empirical evidence, transparency, and openness to rational criticism were traditionally cashed out in terms of the distinction between *disciplines* or *bodies of knowledge*, we attempt now to identify differences that determine whether some *activities* are instances of scientific *research*.

Exercises: (1) Illustrate with a couple of examples how the procedural requirements described above are met by the natural or social science that you are most familiar with.(2) Can you think of other activities that fit these procedural requirements but are not obviously instances of scientific research?

1.4.2. *Functional questions*

It is fairly uncontroversial that the main aim of research is to contribute to a body of knowledge, but not everybody agrees on how to further define the type of knowledge that research is aimed at producing. For instance, in the previous section when we were thinking about potential differences between natural and social sciences, we asked whether the results produced in the course of an investigation of the structure or history of human society can be generalized. One common thought is that, in order for the investigation to count as an instance of genuine scientific research, its results need to be generalizable and that any investigation that does not deliver generalizable results fails to be scientific. Here we shall focus on another requirement, that of novelty.

There seems to be consensus on classifying the confirmation of results and the re-organization of previously known data as research, when confirmation is needed or when there is an element of originality or novelty in the activity. This element of originality might be exhausted by the possibility of drawing further conclusions or making further generalizations from the same body of evidence by re-organizing the data or re-interpreting them in the light of new theoretical assumptions. For Imre Lakatos (1970), who depicts science as the dynamic succession of research programs and not as a collection of theoretical statements, a research program is scientific if it is progressive. In order to be *progressive* with respect to a previous stage of scientific development, the research program must have at least as much empirical content and must be able to offer an explanation for the same phenomena in a way that is at least as satisfactory. In addition, it must make new predictions that can be confirmed by experience. A research program is *degenerating* (i.e. it's still science, but not very good science) if the new predictions that are made are not confirmed by experience.

Although the account by Lakatos has been so far extremely influential within the study of scientific methodology, some problems have been raised concerning the notion of *new* facts and *new* predictions. With respect to what should the facts and the predictions be new? There are different answers in the literature, ranging from temporal novelty to novelty of interpretation. The consequences of which type of novelty we select are very important for the definition of progressive research programs. All that is required by *temporal* novelty is that facts that were not considered likely before can now be predicted. Novelty of *interpretation*, instead, is much weaker and demands just that old facts be revisited and re-evaluated by the research program. The type of novelty required for an activity to count as original research is a difficult issue, and answers might vary depending on the aims of a scientific discipline or even on the stage of development of that discipline. But for the very general purpose of our discussion here, novelty of interpretation seems to be necessary for an activity to be regarded as original research.

 Exercise: Which of the following considerations could give rise to a functional criterion for scientific research – increasing explanatory power, stimulating further debate in one field of investigation, challenging preconceived ideas, having significant technological applications, being compatible with other successful research programs, delivering results in a wide range of interrelated phenomena?

Discuss: What other functional criteria would you add to novelty?

1.4.3. Delimiting research

How can we combine the previous considerations on the functional and procedural dimensions of research to arrive at an account of demarcation? Scientific research is a human activity that aims at contributing to a coherent body of knowledge in a novel way by systematically adopting a critical method. There are two ways in which the function of an activity can be tracked: *subjectively*, looking at the primary intentions of the people engaging in the activity, and *objectively*, looking at what the outcomes of the activity actually contribute to. In many cases these coincide, but sometimes individuals or groups who *aim* at adopting a critical method or at contributing to a coherent body of knowledge in a novel way fail to do so.

The purpose of an activity can contribute to the understanding of the reasons behind human behavior, which is perceived as a legitimate scientific objective, but no critical and transparent method is adopted – some people would say that writing daily horoscopes falls into this category. An activity could be conducted via a critical method without being aimed at extending a body of knowledge in a novel way – for instance the work of a graduate student in physics might be methodologically indistinguishable from the work done by leading researchers in the field, but its main function is to demonstrate the student's competence, rather than to contribute originally to the knowledge shared in the scientific community.

These conceptual resources might also help us make sense of the research-therapy distinction in biomedicine. The same data, obtained via a respectable empirical method, can be used to extend biomedical knowledge or to provide immediate therapy. These functions are non-exclusive and one can argue that obtaining knowledge in biomedicine has always an underlying therapeutic function. Therapeutic attempts using non-validated methods or drugs might generate a hypothesis that could then be tested in a trial.

A good example can be found in the case *Simms v Simms and An NHS Trust* (2002). In this case the court decided that it would be lawful to administer experimental therapy to an incompetent patient, Jonathan Simms, affected by variant CJD. The drug that was administered, *pentosan polysulphate*, had never been used on humans affected by vCJD and was infused directly into the brain, via a risky surgical procedure. The decision of the court was motivated by the serious prognosis of Jonathan Simms and the lack of available alternatives. The case suggests that there are no sharp boundaries between research and therapy and that an activity can have both functions.

1.5.   Good and Bad Science

The distinction between disciplines that are scientific and disciplines that aren't (or activities that count as research and those that don't) does not necessarily carry an evaluative judgment. One might believe that types of investigation that do not meet the procedural criteria for scientific research or are not aimed at contributing to a body of knowledge might nonetheless be extremely valuable and that they should be funded and pursued alongside scientific research. But there are pejorative connotations in the phrases "pseudoscience" and "bad science." A discipline or a type of research that is classed as pseudoscientific typically does not pass the demarcation test, but *presents itself as scientific* or is taken by some sections of the population to have the same status as a scientific discipline or as an instance of scientific research. The negative connotation is due to there being an element of pretence or delusion.

Recently, the debate on creationism has revived the practical concerns that emerge with the necessity to distinguish science from pseudoscience. Is creationism a science? Should it be taught in schools alongside evolutionary theory? The scientific community by and large denies that creationism offers a scientific explanation of the origin of life. The reasons for this position are mainly due to an assessment of procedural factors. Creationism is regarded as a pseudoscience because it introduces supernatural factors in the account of how life developed. Moreover, some of its hypotheses are based on vague notions. For instance, critics argue that the concept of "kind" as one of the different forms in which life was created is not sufficiently elaborated and that this vagueness makes it extremely difficult to point at evidence that would falsify claims about the way in which different kinds interact with each other. These objections to the scientific status of creationism seem to be founded on the view that aspects of the theory compromise both the transparency of the hypotheses put forward and the possibility of their being tested.

The consideration of contextual factors might raise further concerns about the status of creationism. The argument is that there is an alternative discipline that can offer a more empirically founded and overall more convincing explanation of the development of life on earth, evolutionary biology, and evidence gathered in this area is not taken into account or is dismissed too quickly by creationists.

The distinction between good and bad science (or good and bad research) is again different. There are enterprises that, if conducted in a satisfactory way, would contribute to a shared body of knowledge. And there are researchers who subscribe to a methodology that, if applied correctly, would qualify as critical and open to reality-testing. But either the results of the activity fail to make a contribution to a shared body of knowledge or the methodology is not applied correctly. These are cases of *bad* science. The activity does meet *to some extent* the requirements for scientific research, but does not meet them in a satisfactory way and therefore the manner in which it is conducted comes under scrutiny and receives extensive criticism.

In the literature on the history of science there are interesting examples of these failures. Some research programs are aimed at proving the existence of causal agents whose presence can be detected only by experiment. These cases are deviant when the posited existence of these agents is justified with claims that might conflict with experience and when criticism is responded to by offering *ad hoc* adjustments to the initial hypothesis. Another common problem is that the results cannot be reproduced by other researchers or other teams.

The story of the quest for cold nuclear fusion has an episode that is often described as an instance of bad science. In 1989 Fleischmann and Pons who were researching at the University of Utah claimed they had achieved nuclear fusion by liberating deuterium from heavy water (D2O) at an electrode made of palladium. Not discouraged by previous failed attempts at cold fusion, Fleischmann and Pons used a very rudimentary apparatus, essentially a jar of water, whose temperature was measured before and after the experiment. They reported they had found an increase in temperature that they interpreted as a sign that fusion had occurred and that neutrons had been produced. The news spread creating great expectations, especially among the US government who was interested in investigating possible applications of cold fusion for the creation of energy sources.

The two scientists were pressurized into submitting a paper with the results of the experiment for publication and their paper went through peer review and was published. Other teams of scientists reproduced the same experimental conditions but failed to obtain the results Fleischmann and Pons had described, even in those cases in which the equipment used was more sophisticated than the original one. The published paper with the claims that cold fusion had been achieved also encountered a lot of skepticism, as some basic errors in the way the experiment was conducted were revealed, errors that had affected the estimates of heat.

Due to these developments, the claim that cold fusion had been achieved was then rejected and the reaction of the scientific community to the work of Fleischmann and Pons made it to the science pages of major newspapers (e.g. The New York Times). Funding for Fleischmann and Pons's cold fusion project was withdrawn as a consequence of their failure to respond satisfactorily to the challenges by other experts.

In the case of Fleischmann and Pons attempting cold fusion, their activities had all the sociological markers of a scientific activity: they were employed as researchers by a respectable institution and they published their results in a peer-reviewed journal. What went wrong? One style of explanation can focus on factors that are "external" to science as having an undue influence on the way in which the experiment was conducted and the results were published. One might argue that there was excessive pressure from the university to make the results public before similar results were obtained by competing research groups working on similar projects. Another factor is possibly wishful thinking on the part of the investigators: Fleischmann and Pons desired to achieve the much sought-after outcome of cold fusion to the point that they neglected basic methodological issues with their experimental procedure and underestimated the significance of their critics' objections. Another style of explanation could stress what was missing with respect to procedural criteria for healthy instances of scientific research. The claim that cold fusion had been achieved was not confirmed by the available evidence and the experimental results could not be replicated by other teams of researchers.

Whether it is always possible to provide a sharp distinction between cases of bad science and cases of pseudoscience, is up for debate. For instance, is it possible that an investigation aimed at extending knowledge fails so miserably in meeting the procedural criteria that it ceases to be science altogether?

 Exercise: Find another example of bad science and identify the factors that contribute to its being bad.

 Discuss: Do you think that the reasons scientists might have to believe in the truth of their hypothesis can ever be "external" to the practice of science?

Summary
In this chapter we have started framing several questions that might be asked in an attempt to delimit the concepts of science and scientific research. In the following chapters we shall concentrate on distinct areas of the practice of science, among them its methodology, its language, its historical development, its ontological assumptions, its progress, and its relation with the rest of society. In the course of the examination of the philosophical issues emerging in those areas we will find other criteria that might help us come to a more satisfactory account of the distinction between science and non-science.

Here we have approached the demarcation question more generally and reviewed some of the available accounts of the perceived differences between science and ethics; science and metaphysics; science and pseudoscience; natural and social sciences; good and bad science. We have also reflected on the concept of scientific research and its procedural and functional aspects.

The picture emerging so far is a picture of continuity, where perceived differences between what counts as science and what doesn't turn out to be nuanced, and the hopes for a neat and straightforward distinction raised by the criteria of verifiability and falsifiability are defied. But this is not necessarily reason for concern.

Preview of future attractions
In chapter 2 we shall continue discussing what makes science different from non-science by concentrating on the characteristics of scientific reasoning and scientific method. In chapter 5 we shall reconsider the demarcation questions in relation to the rationality of science and the ways it progresses.

Issues to think about

1. Are Marxism and homeopathy pseudosciences? Why?
2. What are the *analogies* between science and metaphysics?
3. To what extent does falsifiability work as a demarcation criterion?
4. What factors determine whether astrology is a science?
5. In what respect do psychology and physics differ?
6. If science is just one tradition of thought among others, what do you think is responsible for the special role it seems to play in contemporary society?

Further resources

An influential and much cited lecture by Imre Lakatos on Science and Pseudoscience can be downloaded from the London School of Economics website (hwww.lse.ac.uk/collections/lakatos/scienceAnd Pseudoscience.htm). In the lecture (first broadcast in 1973) Lakatos summarizes his view on the demarcation problem which can also be found in his *Methodology of Scientific Research Programmes*. Other classical readings include pieces by Popper, Kuhn, and Laudan which can be found in the collection by Curd and Cover (1998).

In more recent literature you will find contributions to the debate whether particular disciplines or research programs are scientifically respectable, rather than answers to the demarcation problem in general. For instance, a number of resources are available on whether research in psychology (Chauvin 1999; Lilienfeld et al. 2004), or intelligent design theories (Haack 2005; Kitcher 1982; Fuller 2007) qualify as having a proper scientific status. For a more comprehensive approach, see two attempts to redefine the demarcation problem in Dupré (1993) and Kitcher (1993).

There are also many fascinating examples of pathological science in two recent publications: Gratzer (2000) and Park (2000).

2 Reasoning

In the attempt to distinguish science from non-science, the nature of scientific reasoning and scientific methodology have always played a fundamental role. Almost all answers to the demarcation question make explicit reference to the way in which scientists reason in order to arrive at the articulation of their theories and at the formulation of the principles that govern natural and social phenomena. Is it legitimate to believe that science has a unique method, rigorous and reliable, which allows scientists to achieve conclusions that are fallible but can be trusted?

In this chapter the main focus will be the nature, strength, and limitations of the so-called "scientific method." Some questions philosophers have been asking, and are still asking, are about the origins of the scientific method, whether it can be applied to the vast and heterogeneous domain of all scientific disciplines and whether it is ultimately reliable. In order to answer these questions, we need a bit of terminology. In the first part of the chapter, we shall learn about different types of statements and arguments. We shall see that there are different ways of reasoning, and that some have the main purpose of acquiring new information and others aim instead at consolidating the knowledge we already have.

In the second part of the chapter, we shall refer to some historical examples to illustrate the important changes that occurred during the revolution that traditionally is taken to mark the birth of mature science. We shall identify the characteristics that distinguish it from previous attempts to understand and describe nature, and we shall see that many factors played a role, including some metaphysical assumptions about the place of humanity in nature; some methodological innovations such as the recognition of the importance of repeatability of experiments and observations; and some institutional changes, as the creation of learned societies and networks of people interested in developing science, which promoted exchanges of ideas and transparency of methods.

The emphasis on innovation should not lead us to think that science was totally re-invented after the Copernican Revolution, or that the described changes were sudden and abrupt. Elements of continuity between Aristotelian and Newtonian physics can be identified, and one feature of science, the use of thought experiments, is an excellent example of this continuity as it spans from Aristotle to Einstein.

By the end of this chapter you will be able to:

- Identify and compare different types of reasoning.
- Recognize a variety of argumentative strategies.
- Assess the strengths and weaknesses of induction.
- Identify some key aspects of the shift in methodology that occurred during the scientific revolution.
- Discuss the role of thought experiments and actual experiments in the practice of science.

2.1. Ways of Reasoning

Philosophers in general, and philosophers of science in particular, think carefully about the many ways we acquire, process and organize information. At any one time, we have a number of different beliefs, for instance that tomorrow it will rain or that $2 + 2 = 4$. That tomorrow it will rain is an *a posteriori* statement, that is, its truth or falsehood depends on how the world will be tomorrow and cannot be established without relying on some form of evidence (sense experience or testimony). That $2 + 2 = 4$ is an *a priori* statement as its truth or falsehood depends on the conventions of mathematics. No evidence can be brought to bear on the truth of the statement and the statement itself does not offer any description of empirical facts.

Another example of an *a priori* statement is "*Either* she will attend the ceremony *or* she won't." We can establish the truth of this statement without relying on any experience, because it is the logical structure of the statement (i.e. the disjunction of two mutually exclusive alternatives) that guarantees its truth.

2.1.1. Justification and truth

How do we form and update our beliefs? We might believe that it will rain tomorrow because we have listened to the weather forecast on the radio. We are likely to maintain this belief if we encounter no evidence against it. We believe that $2 + 2 = 4$ because we were taught that this is the case when we first learned math at school and becoming proficient in these basic calculations was part of the way in which we started understanding the concept of numbers and of operations such as sums and subtractions. We maintain the beliefs we acquired in this way unless we challenge the conventions of mathematics.

By witnessing an accident, watching the thermostat, mixing oil and water in the kitchen, or observing a rare eclipse with our naked eye we acquire beliefs about the behavior of things and people around us. In all these cases, we have *direct* experience of what happens, of what we see or feel, and we gain that experience largely by observation. More often, our beliefs come from indirect sources, some more trustworthy than others (e.g. relevant experts, teachers, books, TV, internet, hearsay, tradition). Finally, there are beliefs that we derive from prior beliefs. If I know that my friend Jacob is allergic to prawns and that in the banquet there is a stir-fry that contains king prawns, then I form the belief that Jacob should avoid the stir-fry.

Our beliefs are justified if we have good reasons to support the content of those beliefs. There is a difference between truth and justification. My belief that tomorrow it will rain can be justified, for instance it can come from a reliable source, and yet it can turn out to be false. Justification does not guarantee truth. On the other hand, I might have a true belief, for instance that my neighbor is a spy, without having any evidence at all to support my belief. My guess or gut feeling in this case happens to be latching onto something true, but my belief is not justified.

There are at least two reasons to care about *justified* beliefs. First, sometimes truth is not enough. Having a justification for the beliefs we have makes it easier to persuade others of the truth of our beliefs. For instance, in the courtroom guesswork and gut feelings do not count. One needs evidence to be able to prove that the defendant is guilty or that no crime took place. Second, justified beliefs are more likely to provide a satisfactory explanation for the phenomena we are interested in, once they are coherently organized. A random set of true beliefs is hardly something we would value, but, if we have grounds for our true beliefs, we are more likely to see the connections between them and draw further implications, thereby potentially extending knowledge.

Sometimes such coherently organized systems of beliefs form theories. We form theories in many different contexts, not just in the context of formal scientific disciplines. We have theories about a variety of things, e.g. how to be successful in job interviews; what started the tension in the Middle East; why John Grisham sells so many books.

Exercises: (1) Think of other examples of (a) unjustified beliefs that turn out to be true and (b) justified beliefs that turn out to be false. (2) Under what conditions would you mistrust (a) direct experience and (b) testimony?

2.1.2. Deductive arguments

Suppose you meet Andrew at a party and he tells you that he likes all the comedies written by Shakespeare. And you reply: "So you must like *Much Ado About Nothing*. It is my favorite!" Your answer shows that you have made a deduction, that is, you have derived a new belief from other two beliefs you had. The structure of the argument is as follows.

Argument 1
Premise 1: Andrew likes all the comedies written by Shakespeare.
Premise 2: *Much Ado About Nothing* is a comedy and it was written by Shakespeare.
Conclusion: Andrew likes "Much Ado About Nothing."

Table 2.1 Logical structure of Argument 1 and of another sample argument.

All As are Bs	All comedies by Shakespeare are liked by Andrew.	All cats are mammals.
C is A	*Much Ado About Nothing* is a comedy by Shakespeare.	Penny is a cat.
Therefore, C is B	*Much Ado About Nothing* is liked by Andrew.	Penny is a mammal.

Let's consider another example of a deductive argument. You need a lift in a hurry and are looking for someone who could drive you to the hospital. Your friend's brother Samir is just across the street, but you don't ask him, because Samir is only 13.

Argument 2
Premise 1: One cannot have a valid driving license at 13 years of age.
Premise 2: Samir is 13 years old.
Conclusion: Samir does not have a valid driving license.

These examples show that we use deduction often, when we derive additional information from the information we already have. In the arguments schematized above, the conclusion does not contain any *novel* information. The conclusion makes explicit something that is already implicitly contained in the premises.

Notice that there is a special relation between the premises and the conclusion. If the premises are true, and the argument is valid, the conclusion cannot be false. One way of describing this relation is to say that the conclusion is *entailed* by the premises, or that it *logically follows* from the premises. The structure of the arguments above is such that the transmission of truth from premises to conclusion is guaranteed.

 Exercise: Identify the logical structure of Argument 2 and find another sample argument that fits that structure.

When we assess a deductive argument, we are interested in validity and soundness. An argument is *valid* if the conclusion logically follows from the premises, as in the examples in table 2. Validity depends on the logical form of the argument. A valid argument is also *sound*, if the premises are true. Argument 3 is an example of a valid deductive argument which is not sound because premise 1 is false.

Argument 3
Premise 1: All Martians are green.
Premise 2: I am not green.
Conclusion: Therefore, I am not a Martian.

2.1.3. Non-deductive arguments

Suppose you have just invited Swati to see the latest Tarantino movie. She shakes her head and says: "I am sorry, I won't come. I have seen three of his movies and they all contained scenes of violence. I don't want to see such a movie tonight." Swati reasoned inductively. She has seen three movies by a director and she has found that they have a feature in common. She expects the latest movie by the director to share that feature. The structure of her argument is as follows.

> **Argument 4**
> *Premise 1*: Tarantino's *Reservoir Dogs* contained scenes of violence.
> *Premise 2*: Tarantino's *Pulp Fiction* contained scenes of violence.
> *Premise 3*: Tarantino's *Jackie Brown* contained scenes of violence.
> *Conclusion*: Tarantino's new movie will also contain scenes of violence.

The premises constitute the *inductive basis* for Swati's conclusion, which is a prediction on the basis of past experience. Let's consider a slightly different scenario. Suppose I have never been to the Southern hemisphere and I have never visited a zoo. If you ask me whether there are any black swans, I might reason as follows.

> **Argument 5**
> *Premise 1*: All the swans I have seen so far are white.
> *Conclusion*: All swans are white.

Here I generalize my belief about the color of swans from a limited sample (constituted by all the swans I have seen in my life) to the whole population of swans. Arguments 4 and 5 are both cases of *enumerative* induction. We reason by enumerative induction when we extend what we have experienced from a number of cases *to the next case* or *to a universal generalization*, and when we extend the relation between two properties in a sample to the relation between those two properties in a population. Argument 4 is a case of enumerative induction to the next case (the next Tarantino movie), whereas Argument 5 is a case of enumerative induction to a universal generalization ("All swans are white"), i.e. a statement of the form "All As are Bs."

By comparing these examples of inductive reasoning to the examples in the previous section, you can appreciate significant differences between induction and deduction. Inductive reasoning is more *daring* than deductive reasoning; the conclusion contains information that does not already appear in the premises. In inductive reasoning we infer from the available evidence to cases that we have not experienced yet (the next case) or that we will never be able to experience in their totality (universal generalization). Our evidence base, no matter how large, can never *guarantee* the truth of the conclusion.

In deductive arguments, if the premises are true and the argument is valid, then the conclusion *must* be true. In inductive arguments, the premises can only *support* the conclusion; making it more likely to be true. What we do not know yet might not resemble what we already know. Australasian swans are black and the next Tarantino movie might not contain scenes of violence after all.

There is another kind of non-deductive inference, called *inference to the best explanation*. Imagine that Dan is a detective investigating a mysterious murder. There are many clues that can lead him to the murderer, but no clue is conclusive. Dan is left with a decision to make and three suspects. He might reason in the following way: which hypothesis best explains all the available clues? Medical doctors often reason this way in order to arrive at a diagnosis. When diagnosing a patient's condition, they might come to the conclusion that the patient suffers from the condition that best explains the occurrence of all the symptoms reported by the patient.

These examples show some of the features of the inference to the best explanation. It is aimed at extending our knowledge rather than making explicit in the conclusion what is already contained in the premises, so it is more akin to induction than to deduction. Inference to the best explanation, as reasoning in enumerative induction, recommends the conclusion as probable given the premises, and not as necessarily following from the premises.

One limitation of inference to the best explanation is that, when a range of viable explanatory hypotheses is selected, the *correct* explanation might not be among the hypotheses under consideration. When the detective or the doctor chooses among viable explanatory hypotheses, they rely on pre-existing beliefs and criteria for what counts as a good explanation that might depend on other theories they accept.

We can talk about an inductive argument or an inference to the best explanation being sound, if both premises and conclusion are true, but we cannot say that they are valid in the same way as a deductive

Table 2.2. Differences and similarities between three common types of argument.

Deduction	Enumerative induction	Inference to the best explanation
The conclusion is entailed by or logically follows from the premises.	The conclusion is made more likely or supported by the premises.	The conclusion is the best available explanation for the truth of the premises.
The content of the conclusion is already implicitly contained in the premises.	The conclusion adds something new to the content of the premises.	The conclusion adds something new to the content of the premises.

argument can be, that is, logically valid. In an inductive argument or an inference to the best explanation the premises do not entail the conclusion. But we can say that an inductive argument or an inference to the best explanation is "correct" if the premises support the conclusion (in this case, the conclusion can still turn out to be false).

 Exercise: Identify three further examples of inductive reasoning or inference to the best explanation.

 Discuss: Does the probability of the conclusion being true vary in the examples? What does it depend upon?

2.1.4.   Reasoning in scientific practice

In the examples of deduction, induction, and inference to the best explanation that we have seen so far, we saw that these ways of reasoning are employed in everyday reasoning and problem-solving. But does the distinction between these types of reasoning matter also to the practice of science?

Deduction is a powerful inferential engine, and it exemplifies the reasoning that takes place when we attempt to prove that the sum of the internal angles in a triangle is 180°; we derive what we want to prove from the axioms of our formal system of geometry and the theorems we can avail ourselves of. But in the empirical sciences, the principles from which we derive further hypotheses to prove are also in need of empirical confirmation, and we use them in our derivations at our own risk. Most philosophers of science would agree that laws of nature do not have a conventional nature, and are not like logical truths or mere definitions of concepts, but they are seen as general statements about phenomena in a particular domain that are themselves typically arrived at via abstraction on the basis of the results of observation and testing. The understanding, accurate description, and prediction of facts that fall under the scope of a theory seem to be hostage to induction. This is to some extent controversial and some philosophers have argued that there is no need for induction in science (see Popper 1953).

Enumerative induction, by grounding beliefs about general phenomena on the past experience of particular phenomena, underpins our everyday actions and expectations and is commonly regarded as the basis of all empirical knowledge. It is a form of reasoning that is *ampliative*, that is, it aims at extending the domain of our knowledge. But it does not provide any certainty, as all generalizations from past observations are fallible.

In science, we have many examples of cases in which what was regarded as a convincing account of the available evidence turns out to be incomplete or inaccurate. Take for instance the debate about the causes of stomach ulcers in medical research. For years researchers believed on the basis of their observations that bacteria could not survive in an acid environment such as the stomach, and therefore ruled out the possibility that bacteria were the cause of ulcers. Stress and spicy food were considered the likely triggers of ulcers and recommended treatment consisted in prescribing drugs that blocked the production of acid. Warren and Marshall recently discovered that there is a bug responsible for many stomach and duodenal ulcers (*Helicobacter pylori*) that lives in the stomach and adapts to its harsh environment. The cure now consists in the elimination of this bug. The world of scientific research is a constant reminder that the hypotheses we have, no matter how well they fit the data previously collected or how entrenched they are in our worldview, may always turn out to be incomplete or inaccurate, and needing revision or replacement.

Inference to the best explanation is allegedly a form of reasoning employed by scientists in circumstances in which the evidence does not rule out conflicting explanatory hypotheses but is better accounted for by some of the available hypotheses than by others. In the *Companion to the Philosophy of Science* Peter Lipton (2001) illustrates this case with the example of Darwin who committed himself to the hypothesis of natural selection because he thought that it provided the best explanation of the available biological evidence.

Another classical example of a scientific discovery made by inference to the best explanation is of Kepler coming to the conclusion that the orbit of Mars is elliptical. According to a controversial reconstruction (Hanson 1958), what Kepler did was to move from the evidence available to him to one hypothesis capable of providing the best explanation for it, on the basis of some background beliefs and methodological principles. There were, at the time, rival accounts for the observations of the motion of Mars. These observations seemed not to support the standard Aristotelian view that celestial bodies move in circles. Some hypotheses maintained that the orbit was circular, but added some devices to explain the irregularities (Ptolemy's equant point device or epicyclical movements). But the hypothesis that the orbit was in fact elliptical had several advantages for Kepler over these contenders: it explained all the available evidence to his satisfaction; it seemed to satisfy the constraints of his astronomical theory; it did not involve unrealistic geometrical devices; and it permitted precise predictions.

Problems with inference to the best explanation emerge when we focus more closely on the suitable notion of *explanation*. Comparisons between the explanatory powers of alternative hypotheses can be made only if we have an objective way to measure those powers, and, as we shall see, philosophers of science offer very different accounts of the conditions that a good explanation has to meet.

The strengths and limitations of inductive reasoning will keep us busy for the rest of the chapter.

2.2. The Scientific Method: Induction

The first philosopher to explicitly state the centrality of inductive reasoning for scientific methodology was Francis Bacon in his *Novum Organum* (New Method). This was supposed to replace the well-established and authoritative methodological text by the Greek philosopher Aristotle, entitled *Organum*. Bacon lived at a time (the seventeenth century) in which the study of the natural sciences was thriving and the authority of the great thinkers of the past was starting to be questioned. He did realize that he was in an important phase of transition, when Aristotelian physics and astronomy were being challenged by the works of Copernicus, Kepler, and Galileo, and with determination he put forward his own vision of how science should proceed.

The inductive method described by Bacon begins with the observation of natural phenomena. He believes that we should present the results of our observations in tables for comparing data. From sensible experience, we then move to lower axioms and from lower axioms to higher axioms, which operate at an increasing level of generality. From the higher axioms we can hope to obtain laws of nature from which new data can be predicted and the observations of new phenomena can be arranged. But Bacon believed that, before data can give rise to axioms, we should use a procedure often called *eliminative* induction. That consists in identifying several explanatory hypotheses for one set of data and ruling out those for which counter-examples can be found. Experiments can be devised to confirm or challenge the hypotheses until one of them survives the tests. The steps described by Bacon form a cycle of knowledge (from low level of generality to high; and then again from high to low) that is supposed to drive scientists closer to the truth.

Apart from the emphasis on induction, there are other important aspects of Bacon's work that we shall look at more closely in the following sections: he endorsed an empirical methodology at a time in which the use of observations and experiments in science was not yet the norm; and he emphasized the collaborative dimension of scientific practice.

2.2.1.   Innovations in the rise of modern science

Bacon is part of the so-called "scientific revolution." The events leading to the acceptance of the Copernican system of planetary motion, culminating with the formulation of Newton's laws of physics, are often characterized as the fundamental chapters in the fascinating story of the birth of modern science. The reason for investing the Copernican Revolution of such a major role in the philosophy of science comes not just from the dramatic overthrow of the Aristotelian and Ptolemaic theories of astronomy and physics, but from the methodological innovations that gradually took place from the end of the sixteenth century.

These are the five most notable changes:

1. the authority of the natural philosophers of the past is challenged on the basis of new observations and insights into the method of science;
2. mathematics comes to be conceived as the language of nature and theories in physics and astronomy are given an explicit mathematical structure;
3. scientists start making use of experiments and mediated observation in a regular and systematic way and actively intervene on nature;
4. collaborative research becomes institutionalized to an extent and gives rise to the development of learned societies;
5. the framework for the explanation of nature (motion, cosmology, physiology etc.) gradually shifts from an *organistic* one, where natural phenomena are seen as the result of intentions, to a *mechanistic* one, where natural phenomena are seen as the effects of causal interactions between the parts of a well-functioning machine, may this be the human body or the entire universe.

These elements are interrelated and taken together give us an idea of how modern science emerged not only from the achievement of theoretical discoveries and the adoption of new methodologies, but also from the gradual establishment of new explanatory frameworks and the emergence of a new conception of reality. We shall focus on a couple of these points and offer some account of the way in which they contributed to transform both the nature of scientific research and the role of the researcher (from "natural philosopher" to "scientist").

There are two ways of conceiving the role of mathematics in the practice of science. One can view mathematics as a mere instrument, intended to facilitate the making of predictions. This was a common approach in a discipline such as astronomy, whose main aims seemed to be the prediction of planetary motion by complex calculations. Astronomers might not have been interested in whether the adopted calculations revealed the physical structure of the universe, as long as planetary motions could be accurately anticipated (*instrumentalist view*). Alternatively, one could regard mathematics as a language that captures the actual relations among observable phenomena and promotes a deeper understanding of those relations (*realist view*).

The main players in the Copernican revolution saw mathematics in realist terms, and used it to obtain an accurate description of reality. The most striking example of the realist conception of mathematics comes from Nicolaus Copernicus himself. In his *De revolutionibus orbium celestium* (On the revolutions of heavenly spheres), he puts forward the idea that the Earth moves and the Sun is in the center of the universe. This was not an unprecedented idea, but had never been supported by such an accurate and detailed mathematical work. Mathematical calculations were not just supporting the heliocentric view, but were the main reason for a quite radical change in the conception of so-called heavenly bodies, and in the conception of the physical world in general, which was taken for granted by ordinary people, intellectuals, and religious authorities alike. What drove Copernicus to the heliocentric model as a representation of the universe was the inadequacy of the Ptolemaic system in delivering accurate astronomical predictions. Mathematics was for Copernicus a guide to reality: if the calculations didn't work, then the physical theory that was supposed to match them had to be replaced.

Analogously, experiments were not unheard of before the sixteenth century and observations in astronomy and terrestrial physics were often conducted and reported in scientific texts. But, again, the role of experimental work did lack the centrality that it has today in our conception of scientific methodology. In the decades leading to Isaac Newton's *Principia Mathematica Philosophiae Naturalis* (Mathematical Principles of Natural Philosophy), published in 1687, the step forward was the legitimization of the use of experiments as an accepted component of the practice of science. In the anatomical and alchemical tradition, it was becoming increasingly common to make and record experiments, for the purposes of teaching and for research. For instance, William Harvey, who discovered how the circulation of blood works, employed experimental techniques and performed vivisection.

The use of instrumentation to aid natural observation was also being introduced, though some innovative instruments were still viewed with suspicion. The most famous example is that of Galileo Galilei's telescope. Galileo constructed a telescope to observe the skies, and following his observations he claimed that the moon had an imperfect surface and that the sun had spots. These were incredibly revolutionary findings, as the received view of the "heavenly" bodies was that they were made of a special substance and that, differently from the Earth, they had a perfectly smooth surface. Many theoretical astronomers doubted the reliability of the telescope, suggesting that it was deceptive and that the observations mediated by such an instrument were not to be regarded as evidence against the accepted view of heavenly bodies. After all, if the moon appeared so much closer to us than it really was through the telescope, what could guarantee that the telescope would not create any other "illusions"? Our answer would be: Galileo's knowledge of optics and his combining lenses in such a way that they would magnify distant objects without otherwise changing their appearance. But if we are reminded that optics was then a very young science and no consensus had been reached on its governing laws, we might begin to understand the struggle scientists had to sustain to justify their use of new equipment and its reliability.

Another important sign of mature science is the recognition of the collaborative nature of research and the need for institutions that promote the exchange of ideas and occasions for discussion. Three important institutions were founded in the seventeenth century: The

Table 2.3 Important figures and stepping stones of the Copernican Revolution.

Copernicus	Argues for the heliocentric system, in which the Earth moves and is not located in the center of the universe.
Kepler	Puts forward two laws of planetary motions: planets move in elliptical orbits and their speed is directly proportional to their proximity to the sun.
Galileo	Argues against Aristotle's theory of terrestrial motions and gathers evidence for the Copernican theory thanks to his observations of the skies conducted via a telescope.
Newton	Combines in his work the efforts of the previous thinkers and builds up a system so comprehensive as to replace both the Aristotelian account of terrestrial motion and the geocentric picture of the universe.

Accademia del Cimento in Italy in 1657, the *Royal Society* in London in 1660, and the *Académie Royale des Sciences* in Paris in 1666. These institutions promoted correspondence among practitioners of science and their publications celebrated the newly established empirical method of science. Some historians have noticed that the increased influence of the experimental method made scientific communication even more important, as scientists had to be able to reproduce experiments in order to assess the results obtained and divulged by others.

Using the example of Robert Boyle's studies in pneumatics conducted via experiments with an air-pump in the seventeenth century, Steven Shapin (1984) claims that in modern science it is no longer sufficient to have the material technology to perform the experiments, but it is also necessary to have the means to let people know about the results of the experiments and establish ground-rules for the acceptance of knowledge claims in science. (This will remind you about the emphasis on transparency and openness to criticism that we considered in the previous chapter as a procedural criterion for the demarcation between science and non-science.) At the time in which Boyle was performing his experiments, the Royal Philosophical Society in London was transforming into a public space where the results of scientific investigations could be shared and disputed. The idea of experimental hypotheses being in need of endorsement by a community of practitioners in order to be legitimized as scientific knowledge was slowly emerging, and now it is a defining feature of scientific research.

Exercise: Can you think of examples of new discoveries in contemporary science that have given rise not just to new theoretical hypotheses but also to methodological innovations?

2.2.2. Thought experiments

When describing the impact of a scientific revolution, it is always easy to highlight the importance of the innovations. But elements of continuity persist together with the ground-breaking methodological achievements. Scientists who engage in experimental work today are part of a community that has mechanisms in place to confirm the reliability of their experimental procedures and the soundness of their experimental results. This apparatus did not come into place at once, and the presence of research institutions was only the beginning of the formation of a scientific community capable of assigning credibility to the work of single researchers or research teams.

Nonetheless, scientists always look for ways of solving the problem of finding successful argumentative strategies to defend new hypotheses and address the skepticism of their audiences. The use of thought experiments, for instance, seems to stand the test of time. As a methodology, it was deployed to provide justification for both Aristotelian and Newtonian physics and was also frequently used in the development of the theory of relativity. There is a long tradition of important thought experiments in the natural sciences, a tradition that starts with Aristotle and Galileo trying to uncover the laws of motion of earthly objects and continuing with Albert Einstein's example of the train struck by lightning that appeared in his first paper on relativity (1905).

Galileo wanted to challenge Aristotle's ideas on falling bodies, and in particular the thought that heavier bodies fall faster. So he imagined what would happen if he threw objects from the Tower of Pisa in order to observe at what speed they fall. If Aristotle had been right, a heavy cannon ball would have fallen faster than a lighter musket ball. But what would happen if we threw a cannon ball attached to a musket ball (i.e. the aggregate of two bodies, one heavier and one lighter)? This new aggregate body is supposed to fall both faster than the cannon ball alone, because heavier, and more slowly, because the lighter part would have slowed down the heavier one. Thus, the line of reasoning based on Aristotle's law of motion leads to a contradiction, which means that the assumption Galileo started with, that heavier bodies fall faster, must be rejected.

 Exercise: Is Galileo presenting here a good argument against Aristotle's law of motion?

Einstein used the famous train thought experiment to argue against the Newtonian view that space and time are absolute. In order to make the point that they are relative instead, Einstein shows that whether events are simultaneous depends on the frame of reference adopted. Imagine that a train carriage is traveling with uniform speed and that a passenger (P) is standing in the middle of the carriage and observing a light bulb emitting rays of light which reach the two extremities of the

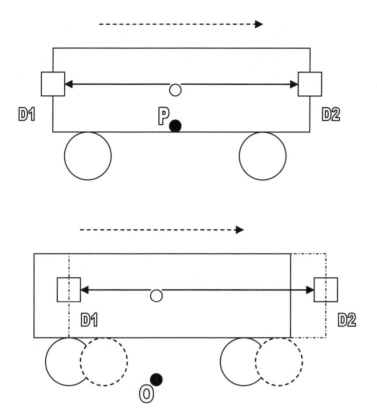

Figure 2.1
Einstein's Train: passenger's perspective

Figure 2.2
Einstein's Train: observer's perspective

carriage where two detectors (D1 and D2) are placed. For this person standing in the carriage, if the bulb is also at the same distance from the two extremities, the rays reach the two detectors at the same time. Now imagine that there is an external observer (O) looking at the train passing by and seeing the bulb and the detectors through an open window. The observer will not perceive the detection of the two rays as simultaneous, because he sees the train as moving. Light will get to one detector (D1) earlier than to the other detector (D2), because although the speed of light is constant, the distance light has to cover to reach D2 is greater than the distance light has to cover to reach D1. The thought experiment is supposed to show that the simultaneity of events depends on the frame of reference (e.g. the perspective of the observers). (See figures 2.1 and 2.2.)

Thought experiments are experiments conducted in the head of the experimenter, rather than in nature or in the lab. The experimenter thinks of a specific situation, imagines something happening in that scenario and draws conclusions on the basis of the consequences of that imagined event. Often the experiment is imagined rather than performed because of physical or technological limitations. At other times, the thought experiment is an intuition-pump used to make explicit our way of thinking about certain situations or to highlight previously undetected inconsistencies.

Whether thought experiments are a valuable scientific resource is open to debate and positions vary considerably. According to Thomas Kuhn (1977, 1979) and Tamar Szabó Gendler (1998) thought experiments can promote conceptual reforms and lead to the replacement of a theory with another. Kuhn claims that engaging in thought experiments does not only clarify the conceptual apparatus within which the scientists operate, but reproduces the clash between conflicting interpretations of nature and therefore prepares the ground for radical theoretical revisions. His view of thought experiments is labeled "constructivist" because instead of challenging an existing theory, thought experiments are seen as challenging a whole way of thinking, an entire conceptual apparatus. His preferred example is the series of thought experiments that Galileo devised in order to show the conceptual inadequacies with Aristotelian physics, in particular the one we described above, expounding the inconsistency of supposing that heavier bodies fall faster but aggregates of bodies fall at intermediate speed.

Not everybody agrees with Kuhn. Some have defended the view that thought experiments are just colorful ways of developing an *a priori* argument and do not have any additional content or significance (see Norton 1996 and Atkinson 2003). Against the constructivist approach, these authors remark that not all thought experiments bring about conceptual revisions. Some for instance are internal to one theory and do not challenge existing concepts. The "empiricist" view defended by John Norton consists in maintaining that only those methods that can legitimately allow us to derive new information from nature (such as actual experiments) can present new information about nature. Given that thought experiments do not pose questions that the observation of nature can answer, they do not extend our knowledge of nature. Similarly to deductive arguments, they just make explicit information that was already available before performing the thought experiment.

Others believe that thought experiments are a way of acquiring knowledge (see Brown 1991 and Bishop 1999). These authors have a "rationalist" approach to the epistemological import of thought experiments, because they concede that thought experiments do not contribute to empirical knowledge about nature, but they maintain that the role of thought experiments is to extend *a priori* knowledge. In particular, replying to Norton, Michael Bishop argues that thought experiments cannot be just arguments, because scientists can come to different conclusions by reflecting on the same thought experiment. This suggests that scientists can use *different arguments* to interpret or reconstruct one thought experiment. James Brown, for example, makes the positive claim that thought experiments *aid* empirical investigations of nature but transcend experience. They are conceived on the basis of intuitions about laws of nature and disclose relations between properties, whether or not these properties are instantiated. Think about the following example, presented by Brown (2004). In his treatise

De Rerum Natura (On the nature of things), Lucretius wants to show that space is infinite. He invites us first to imagine tossing a spear at the boundary of the universe. If it flies through, there's no boundary. If it bounces back, there must be something beyond the boundary. Via this thought experiment we learn something about the *relation* between the *property* of being finite and the *property* of being bounded but we have not empirically demonstrated that space is infinite.

Exercise: Can you think of other thought experiments? How were they used in the context in which you encountered them, to explain a concept, to argue against a position by showing that it generated contradictions or for some other purpose?

Which of the discussed accounts is more convincing? The fact that thought experiments are heavily used by scientists might be a *prima facie* reason against the view that they are "just arguments." This is because of the preconceived idea that arguments that are not based on any new data or novel interpretation of previously available data are not a respectable way of acquiring knowledge in the sciences. This naïve conception of the practice of science rests on the view that decisions about which theory to adopt are made on empirical bases alone. But this is a very simplistic and ultimately misleading way of describing the practice of science. There is greater continuity between science and metaphysics than the logical positivists believed, not only because views about the nature of reality that are formed on the basis of conceptual arguments can promote empirical investigations, as Popper suggested, but also because the aims and the methods of science and those of philosophy sometimes coincide. The use of thought experiments is just an example of this continuity.

Discuss: Is it plausible to claim that the ways in which thought experiments contribute to knowledge in science vary according to the aim of the specific thought experiments?

2.3. The Problem of Induction

Bacon defended the idea that scientific reasoning rests on induction; this idea is still defended by many philosophers of science today. The centrality of induction for forming probable hypotheses about empirical facts has generated a special interest in the justification and rationality of inductive inferences.

According to the traditional account of the *problem of induction*, the philosopher David Hume was concerned with the justification of the inductive inferences that make it possible for us to extend to new cases what we already know on the basis of our experience.

Let's consider the following inductive inference:

1. The sun has risen every day up to now.
2. The sun will rise tomorrow.

Is the step from (1) to (2) justified? We are used to projecting the regularities that we have observed in the past onto the domain of the unknown, but there is nothing to prevent (2) from being false and (1) being true. What grounds the step from (1) to (2)? We could rely on some principle that tells us that the future is going to resemble the past. For instance:

3. Nature operates uniformly.

The statement in (3) is called the Principle of the Uniformity of Nature and can help us support the inferential step from (1) to (2). It is reasonable to assume that the sun will behave as it has always done, because nature repeats itself and the regularities to be observed in nature are not likely to be disrupted. As Bertrand Russell observes in his *Problems of Philosophy*, the conviction that underlies our belief in the Principle of Uniformity of Nature is that everything that has happened, happens now, and will happen in the future is governed by a general rule with no exceptions.

But now, what is the status of (3)? Hume believes that there are three types of statements:

a. statements that can be true *a priori*, independent of experience, such as "Either the sun is rising or it is not rising";
b. statements that we can support by direct observation or other forms of experience, such as "The sun is rising now"; and
c. statements (about the future or about the next unobserved case) that we can support only by inductive reasoning, such as "The sun will rise again tomorrow."

To which category does the principle of the Uniformity of Nature belong? It is an assumption we make, but it is not a logical truth, and it is not true by definition, so we need an *a posteriori* justification for it. But we cannot "observe" the truth of the principle either, so it does not fit either (a) or (b). The only kind of justification the principle can have is an *inductive* justification.

The only justification we have for (3) derives from something like (4):

4. In the past, nature has operated uniformly.

And the inferential step from (4) to (3) is, again, inductive. We generalize from past experience in order to obtain the principle of the Uniformity of Nature. But if we justify all inductive inferences inductively, our justification ends up being circular and we have to concede that inductive inferences cannot be *independently* justified. An argument is circular when, in order to believe that the premises of the argument are true, we already need to assume that the conclusion is true.

Let me offer an example of circularity (loosely drawn from the sci-fi scenario of the *Matrix* movies). Suppose we are asked: "How do you know that your everyday experience is not just the result of an ingenious computer simulation?" One tempting answer is to say: "Because I see that I am surrounded by real objects and real people." But if we did reply that way, we would be begging the question, because in order to accept what we say as true, one has to dismiss the possibility that our everyday experiences are the result of a computer simulation. If our aim is to *show* that we experience objects in the real world, we cannot already *assume* that what we are experiencing is not the effect of a computer simulation.

There have been two main strategies in dealing with the problem of induction: showing that it is not really a problem, but a pseudo-problem created by some conceptual confusion or incoherence; or conceding that there is a problem and suggesting a solution. We shall see just a few examples of each strategy here.

But before we move on to possible responses to the problem of induction, we should notice that, according to recent influential interpretations of Hume (especially Beebee 2006), it is misleading to attribute the problem of induction to Hume, for two reasons. First, Hume was not interested in inductive reasoning *per se*, that is, in the step from observed regularities to a universal generalization or the prediction of a new case, but was interested in reasoning from causes to effects. Second, Hume was not really interested in justifying our way of acquiring knowledge of the empirical world, but was interested in describing the mental mechanism that allows us to form beliefs about the empirical world. So his aim was to address the *psychological* issue about the formation of beliefs about what we have not yet experienced, and not the *epistemological* issue about their justification.

What supports this reading of Hume? First, Hume does provide a psychological explanation of how we get beliefs about the empirical world, by introducing the notion of *habit*, which describes a psychological mechanism by which we observe repeatedly the conjunction of event *A* and event *B* and assume that there is causal relation between them. No attempt is made in Hume's work to offer a justification of the reasoning step from (1) *A* is occurring and (2) *B* will follow from *A*. Second, to conclude that we should be skeptical about induction, or causal reasoning for that matter, would undermine Hume's general philosophical outlook because he is an empiricist and it would be self-defeating for him to compromise the very possibility of obtaining knowledge of the empirical world. This interpretation of Hume can inspire us to view the problem of induction as a pseudo-problem, but in the philosophical literature, and especially within the philosophy of science, the justification of induction has been the object of an endless debate.

 Exercise: Why do we need an independent justification of induction?

2.3.1.   Can we dissolve the problem of induction?

Peter Strawson (1952) is not at all convinced that there is a "problem" with induction. He argues that our concern about justifying induction arises from the attempt to answer a senseless question: is it *reasonable* to believe in induction? Strawson claims that the meaning and use of the word "reasonable" already presupposes our conformity to inductive standards. We would not know what is reasonable to believe unless we could rely on inductive inferences.

Strawson diagnoses the problem of induction as the hopeless search for a justification of induction that meets the standards of deductive reasoning. What philosophers seem to require, according to Strawson, is a deductive argument entailing that inductive arguments are sound. But it is not clear that this is a legitimate project. The only legitimate question that can be asked about the justification of induction, according to Strawson, is whether the evidence provided in specific inductive arguments supports the conclusion of those arguments. The answer to this question cannot be about induction in general. There are good inductive arguments, in which the available evidential basis is sufficient to make the conclusion probable, and bad inductive arguments in which the conclusion is not made probable by the available evidential basis.

Not satisfied with Strawson's defense of induction, Max Black (1954) also contributed to this debate. For Black as well as for Strawson, the problem of induction is not really a problem. The mistake that philosophers make, according to Black, is to misinterpret the claim that there cannot be an inductive justification of induction. The reason why the claim is upheld is to be found in the alleged circularity of any inductive justifications of induction, but Black argues that there is a legitimate and non-circular way of justifying induction *inductively*. He starts by distinguishing two levels of discourse, the first level, which concerns objects, properties, and relations in the world, and the second level, which concerns arguments and logical rules.

At the first level, we can say that the fact that all the swans observed so far are white supports the claim that all swans are white. This argument is justified by an inductive rule R (second level) according to which we can argue from "all examined instances of A have been B" to "All As are Bs." The rule justifies the argument, but what justifies the rule? According to Black, there is an inductive argument that can be used to justify the rule and it goes more or less like this: R has been reliable in all the examined instances of the use of the rule, so we have good grounds to believe that R will be reliable now.

The question at this point is: have we incurred any circularity? We have justified one inductive argument with the inductive rule and justified the inductive rule with *another* inductive argument. According to Black, there is no circularity involved. An argument is circular if one of the premises is identical to the conclusion or if the premises are such that we would be in no position to know them unless we already knew the conclusion. The argument used by Black to justify induction is not circular according to this definition of circularity.

But there is another worry lurking in the background. Maybe the argument put forward by Black does not qualify as strictly circular, but is at risk of *infinite regress*. That is, the justification of the inductive argument regarding white swans is indefinitely deferred as it depends on the justification of the rule of inference, which in turn depends on the justification of another inductive argument, which in turn will depend on the justification of the same rule or another inductive rule, and so on.

Elliot Sober (1988) finds the project of providing a justification of induction by defending the assumption of the uniformity of nature misguided. He notices that how nature behaved in the past does not seem to be a good reason to trust induction as opposed to *counter-induction*. Imagine that there is a method of inference (called counter-induction), according to which nature is not uniform and past regularities are not likely to remain the same in the future. Now it is easy to construct a counter-inductive justification of counter-induction based on the uniformity of nature. Counter-induction was unreliable in the past, because nature has proved to be mostly uniform, so we can expect it to be reliable in the future.

But Sober does not think that the principle of the uniformity of nature, in its Humean formulation, is plausible. And this compromises the plausibility of the standard formulation of Hume's problem of induction. The claim that nature is uniform is too vague to be of any use, and, in that form, it cannot be an assumption that we want to hold on to when we make inferences. In some respects nature is uniform but in other respects nature is not. Sober's example is that of color: we expect the color of all emeralds to be green, always, but do not expect the color of leaves to be always green. Lack of details in the formulation is not the only problem with the principle of uniformity of nature: according to Sober, attempting to find one assumption about the nature of things that grounds all of our inductive inferences does not coexist happily with the flexibility of our inductive practices. It is more plausible that some background beliefs are indeed necessary for inductive inferences to succeed, but which beliefs support which inferences depends on the particular inferences.

 Discuss: Which one of the above dissolutions of the problem of induction seems more plausible to you?

2.3.2. *Attempting a solution to the problem of induction*

Wesley Salmon (1974), inspired by a previous analysis by Hans Reichenbach, argues that both deductive and inductive ways of justifying induction are doomed to failure and explores a more pragmatic alternative. What if we can show that trusting induction is the only rational option? Salmon claims that it is more prudent to bet on the success of inductive reasoning rather than bet on its failure.

Let's start from the assumption that either nature is uniform or it is not and that, if we decide not to use induction, the available strategies for the prediction of unknown events are non-inductive (e.g. reading tea leaves, crystal-gazing or wild guessing). If we use non-inductive strategies and nature is uniform, it is likely that we fail in our predictions. If we use non-inductive strategies and nature is not uniform, it is *still* likely that we fail in our predictions. If we use induction instead and nature is not uniform, it is likely that we fail in our predictions, again. But, if we use induction and nature is uniform, we are likely to succeed in our predictions.

Although induction does not *guarantee* success, as we have no way of independently finding out whether nature is uniform, induction still works better than the available alternatives. This is a way of arguing that it is rational to use induction, but does not answer the question whether inductive inferences can be independently *justified*. So this attempted solution misses the point (or, on a more charitable reading, provides a redefinition of the problem of induction).

In *Objective Knowledge*, Karl Popper claims to have solved the problem of induction. First he reformulates the problem by asking the following question: "Can the claim that a universal theory is true be justified on the basis of the truth of certain observation statements?" If this is the question, then Hume was right to think that the answer needs to be "no." There is no way of justifying a *universal* theory by relying on the truth of some of its instances that can be confirmed by observation. But the solution comes from asking a similar question, which has a positive answer: "Can the claim that a universal theory is *false* be justified on the basis of the falsity of certain observation statements?"

According to Popper, we need to give up the idea that science is based on inductive inferences and accept a form of *deductivism*, which works in the following way. First, we formulate a new hypothesis. Then we derive from it via deduction some statements whose truth or falsity can be ascertained by observation. If the statements are found to be true after careful testing, we haven't shown anything conclusive about the hypothesis, which needs to undergo further testing. If the statements are found to be false, then we have shown that the hypothesis is false (as it yields the wrong empirical consequences) and have made some progress. We have, in other words, *falsified* the theory. There are problems with Popper's suggestion, which we will encounter when we address issues in theory confirmation, but it is interesting to note here that Popper thought that science does not, and should not, rely on inductive inferences and suggested what he viewed as a plausible alternative.

 Exercise: Think about Popper's solution to the problem of induction and answer the following questions: (a) How do we arrive at new

hypotheses if not by induction? (b) Can we ever be justified in accepting a theory by using Popper's method of testing?

 Discuss: Do you think the problem of the missing justification of induction can be solved?

Summary

In this chapter we have first defined three ways of obtaining or consolidating knowledge that correspond to three different rules of inference: deduction, enumerative induction, and inference to the best explanation. We emphasized the importance of inductive inferences for the practice of science and discussed some attempts to justify the reliability of induction.

We described some of the methodological innovations that have characterized the scientific revolution and regarded them as important in our contemporary understanding of science: the acceptance of inductive reasoning as the basis for all empirical sciences was one of them. Some of the other aspects were the use of mediated observation and experiments in the practice of science and the foundation of institutions that promoted the development of a community of practitioners in constant dialogue with each other.

All these methodological features of the practice of science could be regarded as criteria to distinguish scientific activities *proper* from pseudoscience, but this move would be too hasty. First, there are elements of continuity between science before and after the Copernican revolution, including the role of metaphysical assumptions driving research projects. These elements should make us wary of the phrases "pre-science" or "primitive science" to refer to the attempts to describe the physical world made by Ancient Greeks and medieval natural philosophers. Second, we should keep in mind that there are significant methodological differences in the practice of different scientific disciplines. The evidence is not gathered in biology in the same way in which it is gathered in physics and very few generalizations about method can be comfortably made. Third, the continuity of argumentative strategies and aims seems to reduce the divide between "mature" and "immature" science. Scientists still want to explain the phenomena around us in a systematic and satisfactory way and still use good old-fashioned arguments and thought experiments to do so.

Preview of future attractions

In chapter 3 we shall continue discussing the role of induction in the practice of science and assess different strategies for the confirmation of scientific theories. We shall examine Goodman's riddle of induction and introduce another paradox of theory confirmation. We shall explore further the notions of direct observation and mediated observation and discuss the relation between theory and observation in relation to the structure and formation of theories (chapter 3), the nature of theoretical terms, and the realism/anti-realism debate (chapter 4), and the nature of progress in scientific change (chapter 5).

Issues to think about

1. Can one be a scientist without performing any actual experiment?
2. In what ways is collaboration among scientists beneficial to the acquisition or consolidation of knowledge?
3. Do thought experiments really contribute to the making of science?
4. Which type of reasoning is most likely to guide scientific practice?
5. Why are philosophers worried about the justification of induction?
6. In what ways is Bacon's inductivism still a good approach to scientific methodology? In what ways is it outdated?

Further resources

For an introduction to the problem of induction and the traditional debate, read some of Hume's *Treatise* (book I part III) and his *Enquiry* (sections IV and V), Popper (1953, 1974) and Russell (1967, chapter 6). An accessible and thought-provoking interpretation of Hume's problem of induction can be found in Beebee (2006) and Popper's deductivism is discussed in detail in Newton-Smith (1981, chapter 3).

The debate about the nature of thought experiments between Norton (2004) and Brown (2004) will make you reflect further about the use of thought experiments in science and is a good source of examples.

If you want to know more about the Scientific Revolution or other significant episodes in the history of science in general, a guide to further reading is available at the site of the History of Science Society (www.hssonline.org/). You will also find useful information about new publications and conferences.

3 Knowledge

Here we shall continue with the investigation of epistemological and methodological aspects of science from the previous chapter. Our focus will be the structure, formation, confirmation, and explanatory role of scientific theories.

In the last chapter, we saw the first attempts to characterize the scientific method in Bacon's work. The simple inductivist picture of the practice of science that we have briefly introduced emphasizes the *empirical* foundation of scientific theories. Scientists empty their minds of any preconceived opinion and *open their eyes*: they collect data on the basis of observations and experiments. They abstract from the results of these observations and experiments, and formulate hypotheses of increasing generality. Then they test the predictions they can make on the basis of these hypotheses and conduct more observations and experiments. The idea is that by inductive reasoning we can arrive at a generalization about a type of object having a certain property in a certain context, if objects of that type have been found to have that property in relevantly similar contexts.

This picture of how scientists operate is grounded on the reliability of inductive inferences and takes observations and experiments to be the building blocks of theories. But the relation between theory and observation needs to be explored further. As Popper did, some object to the inductivist picture of science that the stage of observation cannot be prior to and independent of the formation of a specific hypothesis. When scientists observe, they always have some expectations to guide and frame their observations, some idea of what they are going to see.

Our scientific theories make sense of the observations we make in a systematic way and one of the purposes of scientific theorizing is the prediction of future events. But many of the observations would not be even conducted if not to test a particular hypothesis, or a set of hypotheses, and the observations that constitute confirming evidence for our hypotheses are often mediated by sophisticated equipment whose reliability depends on further theoretical assumptions. Some philosophers talk about the *theory-ladenness* of observation, that is, the fact that the data we acquire through direct or mediated observation are not neutral with respect to all theories, even prior to being interpreted.

Theories explain and predict events, but how do they do so? We shall review different models of theory confirmation and explanation and present the two main conceptions of scientific theories: the *syntactic* and *semantic* conception. Both conceptions seem to face challenges coming from the theories of confirmation and explanation.

In the course of our brief introduction to the notion of theory confirmation, we shall see that it is difficult to characterize the way in which observations lend support to scientific hypotheses. When we turn to different models of explanation, we shall review the conditions that make scientific theories able to provide adequate explanations of the phenomena that interest us. Can we explain a fact without appealing to a law of nature? How do we decide between competing explanatory hypotheses of the same phenomena?

By the end of this chapter, you will be able to:

- Discuss and assess different philosophical accounts of the formation and nature of scientific theories.
- Provide an account of the relation between theory and observation.
- Discuss and assess different philosophical accounts of the confirmation of scientific theories.
- Discuss and assess different models of scientific explanation.
- Form an opinion about what makes a set of statements a *scientific* theory.

3.1.   What is a Theory?

What we expect from a theory in a specific domain is a coherent and systematic explanation of why some facts occur in the way they do and a reasoned way of predicting the facts that will occur in the future. For instance, a theory about the motion of terrestrial objects tells us on the basis of which principles objects of a certain size move in the way they do and also allows us to predict how they will move in the future or how they would move in non-actual contexts (e.g. if there were no inertia).

Any scientific theory involves a number of statements, ranging from empirical statements about particular phenomena that we can observe (either with our naked eye or by using scientific instruments), and values that we can measure, to general principles. In physics, an example of an empirical statement about a particular event is: "The body is falling at the speed of 40 km/h." An example of a general principle is the principle of inertia, according to which a body will preserve a constant velocity unless acted upon by an unbalanced force. In social psychology an example of an empirical statement about a particular event is: "After performing a boring task as part of a psychological experiment, people who were offered no incentive to participate rated the task positively." An example of a principle is the cognitive dissonance hypothesis: people attempt to reduce the conflict between attitudes they are aware of (e.g. beliefs, decisions, preferences, emotions) and change their behavior accordingly.

In the natural as in the social sciences, principles can be confirmed on the basis of empirical statements about particular phenomena which belong to the domain over which the principles range. When we ask how scientists *form* theories, we are often interested in the way they proceed from observational statements about particular events, agents, objects and so on to general principles, and ultimately laws. Philosophers of science have long attempted to provide a reconstruction of what scientific theories are and how the transition between the observation of facts and the formulation of general hypotheses or principles can be described.

 Exercise: Can you think of other generalizations in the natural and social sciences? What do they all have in common?

3.1.1. Conceptions of scientific theories

According to a popular version of the *syntactic* conception of scientific theories (from *syntax*, the study of the rules that determine how sentences are formed), theories are collections of statements that can have a formal representation as axiomatic systems. The fundamental idea is that we can separate the logical structure of the theory (*calculus*) from its factual content (Carnap 1967; Hempel 1970). The un-interpreted sentences are connected to each other logically: e.g. from the axioms theorems are derived by deduction. When axioms and theorems are interpreted, then we obtain the statements that compose the body of the theory. Such statements contain logical, observational, and theoretical terms. Theoretical terms can be given meaning by correlation with observational terms via correspondence rules.

Logical terms are terms such as "and" or "or" that signify a logical relation between predicates or propositions. In the sentence "I have an old car *and* a new mountain-bike," the term "and" serves to express the relation between my having an old car and my having a new mountain-bike (which is a relation of *conjunction*). The sentence expresses a true proposition if it is both true that I have an old car and that I have a new mountain-bike.

Observational terms are terms which we can apply through direct experience. The term "old" in the previous sentence is observational, because I can establish through direct observation and experience that my car is old (e.g. by looking at it, hearing the noise produced by the engine, observing that it is slow at starting up when the weather is cold).

Theoretical terms are those terms that are neither logical nor observational. On example could be the term "diabetes." We cannot establish whether someone has this condition by mere un-aided observation. In order to see whether a person has the characteristics typical of this condition further investigation is necessary: adequate tests need to be made and interpreted by a medical doctor. The method of finding out whether someone has diabetes can be broken down in a series of observations, so it is in principle possible to establish a correspondence between statements about diabetes and statements containing exclusively observational terms.

> *Exercise: Think of other examples of observational and theoretical terms. Can you think of a term that is observational in some contexts and theoretical in others?*

The attraction of the syntactic conception is that one can in principle separate the logical structure or skeleton of a theory (constituted by the un-interpreted axioms and the theorems) from its empirical content and meaning. But the difficulty in drawing a sharp line between theoretical and observational terms, together with the problem of specifying satisfactory correspondence rules by which theoretical terms get their meaning, has led philosophers to develop an alternative account of scientific theories, the so-called *semantic* conception.

There are different versions of the semantic conception (Van Fraassen 1980; Giere 1988; Suppe 1989), but they all share the rejection of the syntactic approach. A scientific theory cannot be adequately presented as a formal axiomatic system written in the language of logic that is, at a later stage, subject to semantic interpretation. Rather, any theory should be presented as a set of theoretical definitions and a set of statements claiming that several things in the world satisfy those definitions (*theoretical hypotheses*). In this account, it is not possible to hypothesize a sharp divide between logical structure and meaning, as definitions are not necessarily expressed in a formal language. Moreover, the relation between the statements of a theory and the world of experience is no longer hostage to the identification of correspondence rules, but relies on the scientists' creation of abstract *replicas* or *models* of reality that fit the provided theoretical definitions.

Let's return to the examples mentioned above and see how statements of a theory could be arrived at and tested in the two conceptions we presented. For the syntacticist, the principle of inertia which can be derived inductively from the observation of how bodies move would be formulated as a logical principle, and work as an axiom in a system. From the axioms theorems would be derived and interpreted so as to contain theoretical and observational terms as well as logical terms. They would become statements about, for instance, the way in which bodies of a certain mass would move under the influence of certain forces in an environment where some variables were controlled. The theoretical terms contained in those statements (e.g. "inertia," "force," "acceleration," etc.) would receive their meaning on the basis of their correlation with observational terms. Those statements would then be ready to receive further empirical support by relevant observations and experiments aimed to verify (or falsify) them. The empirical support for those statements would transmit empirical justification to the principles from which they are derived.

For the semanticist, there would be no need to translate scientific statements into the sentences of a formal language at the time of axiomatization, and then reverse the process for the purpose of testing the derived theorems in the light of observation and experiments. The principle of inertia would be formulated on the basis of inductive evidence, as we saw before, and then inertia would be defined. A theoretical hypothesis that satisfies that definition would be put forward. Such a hypothesis would refer to an abstract and often suitably simplified replica of reality (an environment in which some variables would be better controlled) that could serve as an idealized model in which to observe the basic relations between, say, the velocity of a body falling and the (amount of) force impressed upon that object. There would be no need to go through the steps of interpretation and work out the correspondence rules for all the theoretical terms deployed, because the definition and the hypotheses would be already "interpreted," that is, expressed in the concrete language of science, and not in the abstract language of logic.

Frederick Suppe (1989) insists that semanticists accurately describe how scientists proceed. Models and abstract replicas are created to illustrate the complex relations among properties. Let's consider two examples. In psychology, behaviorism aims to identify the parameters that lead people to behave the way they do, and describes behavior as the function of stimulus and response patterns. As it is difficult to isolate motivational patterns in humans, given the number of interests they have, the relation between stimulus and response can be demonstrated by observing the behavior of other minded beings, whose interests can be identified and controlled more easily. In the experimental setting a hungry rat presses a lever and obtains food. When the same situation repeats itself, that is, the rat is hungry again, the lever will be pressed again and food will be expected. By themselves, these parameters are not sufficient to explain the variety and complexity of human or even animal behavior, but they provide a model that approximates the reality one wants to investigate by allowing scientists to explore the relation between stimulus and response.

In physics, the Bohr model of the atom (so-called as it was developed by Niels Bohr, in 1915) describes electrons as circling around the atomic nucleus (composed by neutrons and protons) in the same way in which planets circle around the sun. This model is useful for many educational and other explanatory purposes, but does not provide an accurate description of the nature of the atom, as the relation between the nucleus of the atom and the orbiting of electrons is dissimilar from the relation between the sun and the orbiting of planets. For instance, in the solar system planetary orbits are confined to a plane, which is not true of the orbits on which electrons travel, and the attractive force that keeps electrons orbiting around the atomic nucleus is much greater than the gravitational force acting on the planets of our universe.

To find examples of models in science is not hard, but to account for the way in which models contribute to the formation and the development of scientific theories is another matter. Consider the following questions:

a. What is the relation between the model and the phenomena to be understood? Do they need to have the same structure?
b. What is the relation between theory and relevant models? Can models *replace* theories, or do they just *supplement* them?
c. What are models? Are they physical entities or fictions?

To some extent, the plausibility of the answers one could offer to questions (a) to (c) depends on the type of model under consideration, so one might argue for a pluralist approach to the function and nature of models in science. But depending on some of the answers provided to the questions above, different versions of the semantic conception of scientific theories have been formulated and defended. Let me offer a sketchy example of one of these debates.

Consider the model (e.g. planets orbiting around the sun) as a representation of a phenomenon (e.g. electrons orbiting around the atomic nucleus). What is the relation between the phenomenon represented and its representation? Some argue that there must be isomorphism between the two, where "isomorphism" means literally "sameness of structure" (Van Fraassen 1980; Suppe 2002). Others argue that it is sufficient to establish a relation of similarity between the representation and the phenomenon represented (Giere 2004; Teller 2001). It has been observed that the latter version is more promising as a general account of the relation, as it can accommodate models which are inexact because they simplify too much. On the other hand, they observe that the account is too vague to be genuinely useful if no degrees of similarity are specified (see Frigg and Hartmann 2006).

 Exercise: Find another example of a model used in science and revisit questions (a)–(c) above in the light of the new example.

 Discuss: In what way is the model useful? What are the similarities and dissimilarities between the phenomenon represented and its representation?

We have presented two alternative conceptions of the nature of scientific theories. But Ronald Giere (2000) has argued that the debate between syntactic and semantic theory is out of date. The motivation for discussing the nature of scientific theories made sense in the context of attempting to offer a philosophical reconstruction of science, which is no longer firmly on the agenda. To some extent, the emergence of different types of legitimate scientific disciplines in which theories do not rely on mathematical models as much as physics does (such as biology and psychology), has shown that it is unrealistic to try and describe the structure of *all* scientific theories.

Moreover, the debate between semanticists and syntacticists does not exhaust the *spectrum* of possibilities. There are alternative accounts that can be found in the work of Kuhn and Feyerabend, who have been interpreted as developing a historicist view of scientific theories (see chapter 5 for more details); and in the work of Thagard who advocates a computational account of scientific theories. These authors highlight the importance of finding solutions to pressing problems in the actual practice of science and they observe how the procedures that count as a rational way to promote the progress of science might vary according to the historical and wider cultural context in which scientists operate.

3.1.2. The hidden complexity of observation

In order to assess inductivism and the conceptions of theories we described above, we need a better understanding of the relation between *observing* and *interpreting* an event in the light of some theory.

What counts as observable? Rudolf Carnap (1966) claims that the scientist does not use the term "observable" to refer to properties that can be directly observed, but to properties that can be detected through our senses, such as being red or blue, hot or cold, smooth or rough. According to this definition of what is observable, "warm" qualifies as an observational term, but "electrically charged" does not. Suppe (1989) agrees that there are contexts in which in-principle observable properties cannot be ascribed by relying on observation. For instance, we said that "warm" is a paradigmatic observational term, but we could not establish by direct observation whether the sun has this property. And there are contexts in which the presence of more abstract properties such as being electrically charged can be verified by simple observation (e.g. what happens if we stick a finger in a socket).

The importance of observation being *direct* or *immediate* is also controversial. If we do not accept that the entities we see through a microscope have the same status as those we see with our naked eye, what are we going to say about what we see through our reading glasses? There seems to be a strong argument for the view that the distinction between directly observable and unobservable is not sharp but takes the shape of a "continuous transition" (Maxwell 1962).

Bas Van Fraassen (1980) does not find this *continuity argument* convincing. Even if we agree that the dichotomy between observable and non-observable has some elements of arbitrariness in it, there are two cases that must be distinguished. Some objects that are seen via the mediation of an instrument can also be seen with the naked eye in the appropriate conditions (e.g. astronauts can have a good look at Jupiter's moons without a telescope). But there are objects which can *never* be seen directly, such as a blood platelet, and the impossibility is due to our limitations as human beings. In this latter case, instruments are necessary to the experience of objects, and this difference might have an important role to play in the debate about the ontological status of these objects (which we will come back to in chapter 4).

 Discuss: Is it really impossible to overcome current human limitations and see blood platelets without the assistance of instrumentation?

Ian Hacking (1981) draws some interesting conclusions from the history of microscopes and challenges the distinction Van Fraassen makes between what is possible and impossible for humans to see. Maybe the astronaut gets a good view of Jupiter's moons by flying in space, but the microscopist gets a good view of platelets by relying on the map of interactions between the specimen and its image (if the map is well-made). Hacking convincingly argues against the notion of observation as something passive, that depends on the features of the property or object to be observed, and suggests instead that observation is a *skill*, and it involves some *doing*. Looking by itself does not allow people to use microscopes effectively: it is practice that gives people the skill to *see through* (or better, *with*) a microscope and this ability does not necessarily depend on the adoption of a particular theory (although the adoption of a particular theory is necessary to *build* a microscope).

 Exercise: In what ways is seeing through a microscope different from seeing through a pair of spectacles with colored lenses?

Discuss: What do you think about the "continuity thesis"? Is there a sharp distinction between what we can and cannot observe? Consider, for instance, the example of magnetic resonance imaging (MRI), a technology used to evaluate tumors or examine suspected brain damage. Is the tumor or the brain damage observable?

3.2.   Theory Confirmation

It might still be open to debate what scientific theories are, and how to best account for the relation between observations and theoretical hypotheses, but philosophers of science agree that theories can be confirmed or disconfirmed by further observations and experiments, and this guarantees their empirical grounding.

By the claim that an observation confirms (or disconfirms) a hypothesis philosophers mean that the observation constitutes confirming (or disconfirming) evidence for that hypothesis. As we saw, universal hypotheses cannot be verified once and for all, no matter how many observations are in accord with them, as the number of these observations will always be finite; and an existential hypothesis cannot be falsified once and for all, because an object whose existence has not been observed yet could be discovered in the future.

In this section we shall look more closely at the way in which a theory can be confirmed by the available evidence and examine some of the puzzles that have generated new ways of articulating the notion of confirmation. Whereas Carl Hempel aims at accounting for confirmation as a purely logical relation (*hypothetical-deductive model*), other philosophers claim that only by introducing the notion of probability we can capture those aspects of confirmation that are relevant to actual scientific practice (*Bayesian approach*).

3.2.1. The paradox of the ravens

Although the idea of a hypothesis being confirmed by observation is intuitive and straight-forward, it is hard to offer a precise characterization of the relationship between the empirical evidence and those statements in a theory that the evidence is supposed to lend support to. The starting point for the debate is what is known in the literature as *Nicod's Condition*: a hypothesis of the form "*A* entails *B*" is confirmed whenever we observe the presence of *B* in a case of *A*. Thus, confirmation is the relation between a hypothesis and an object or event.

Let's stick to the classic example. The following hypothesis (U) "For all instances of x, if x is a raven, then x is black" will be confirmed by the observation of a black thing which is a raven. The idea behind this condition is that when you have a universal statement, the probability of its being true is raised by finding one instance of the generalization. A black raven confirms (U) and a non-black raven disconfirms it.

Hempel (1945) develops his own account of theory confirmation on the basis of Nicod's Condition. Hempel embraces the idea that confirmation concerns the relation between hypotheses and observations, but conceives it as a *logical* relation between *statements*, analogous to that of logical consequence. An observation *statement* acts as confirming or disconfirming evidence for a statement reporting a scientific hypothesis. Hempel is not satisfied with the details of Nicod's account, as it violates the principle of logical equivalence. This principle states that, if two hypothesis-statements (H1 and H2) are logically equivalent, then any observation-statement (O1) that confirms H1 must also confirm H2. (U1) "For all instances of x, if x is a non-black raven, then x is a raven and is not a raven" is logically equivalent to (U) "For all instances of x, if x is a raven, then x is black." But no observation can confirm (U1), because nothing can be a raven and not a raven. That means that the principle of logical equivalence is violated by Nicod's condition.

This is a problem for Hempel, because he doesn't want to develop a notion of confirmation according to which the relation between the observed phenomenon and the hypothesis made more or less probable by that phenomenon is hostage to the ways in which hypotheses and observation statements are formulated. The observation statement

Table 3.1 A paradox for Hempel's theory of confirmation.

The observation statement "This is a white shoe," which confirms
H2, also confirms H1 because H1 and H2 are logically equivalent.

H1 All ravens are black. $(x)(Rx \rightarrow Bx)$	This is a raven and it is black. (Ra and Ba)
H2 All non-black things are non-ravens. $(x)(\sim Bx \rightarrow \sim Rx)$	This is a non-black non-raven. ($\sim$Bb and $\sim$Rb)

"This is a raven and is black" is confirming evidence for the statement
(U) "For all instances of x, if x is a raven, then x is black," and the relation between them is analogous to that of logical consequence. But also
the observation statement "This is not a raven and is not black"
confirms (U) "For all instances of x, if x is a raven, then x is black"
because it confirms a statement logically equivalent to (U), that is (U2)
"For all instances of x, if x is not black, then x is not a raven."

But defending the principle of logical equivalence raises other problems, which Hempel does acknowledge. It seems very counter-intuitive,
given our commonsense notion of confirmation, to argue that, if we find
something which is not black and is not a raven, let's say a white shoe or
a green leaf, the observation statements "This is a shoe and it is white"
and "This is a leaf and it is green" support the universal statement that
"For all instances of x, if x is a raven, then x is black." This is puzzling
because the observation of a white shoe or a green leaf seems to bear no
relevance whatsoever to the hypothesis that all ravens are black.

Hempel first considers some potential solutions to this problem,
including the introduction of a qualification determining the field of
application of each universal hypothesis. For instance, the hypothesis
"For all instances of x, if x is a raven, then x is black" would only be
assessed against the class of ravens and the observation of shoes and
leaves would not be deemed as relevant to its assessment. But this
measure would introduce an element of arbitrariness that is difficult to
justify and that would seem to be an *ad hoc* expedient. Instead, Hempel
wants to endorse the counter-intuitive conclusion: he argues that, even
though the claim that the observation of a white shoe confirms that all
ravens are black appears puzzling, it should be embraced. The reason
for our initial puzzlement is that we are used to conceiving of a universal hypothesis as establishing some truth in relation to a class of individuals, whereas universal hypotheses should be regarded as issuing a
prohibition that applies to all objects: If something is a raven, then it
cannot be other than black.

In confirmation, according to Hempel, we are not just assessing the
relation between some given evidence and the hypothesis, but we

are assessing the hypothesis against the *conjunction* of the new evidence we have just gathered *and* all the evidence previously available to us. The observation of a white shoe does *weakly* increase the probability that all ravens are black. With this account of confirmation, tacitly taking into account prior evidence, another counter-intuitive conclusion is embraced: the observation of a white shoe provides confirming evidence not only for the statement that all ravens are black, but also for the statement that all ravens are green. How can the observation of the very same object – a white shoe – confirm that all ravens are black and that all ravens are green, at the same time? The consideration of previously gathered evidence might help answer this challenge, but the revised account leaves Hempel's critics unsatisfied.

Exercises: (1) Apply the principle of logical equivalence to another example of confirmation. (2) Can you spot other undesirable consequences of Hempel's account of confirmation?

3.2.2. *Alternative accounts of confirmation*

Wesley Salmon (1975) finds that Hempel's defense of the hypothetical-deductive model of confirmation is based on a fundamental confusion between two conceptions of confirmation: on the one hand, we can say that a theory is accepted by a scientific community because it has received confirmation from the available evidence; on the other hand, we can say that a specific observation increases the probability of a theory, independent of the other available evidence. This distinction does matter, as we can appreciate in the following case: a hypothesis with a low degree of probability given the available evidence can be "confirmed" (in the latter sense) by one additional piece of evidence, and yet its probability with respect to the conjunction of prior evidence with new evidence has not increased. If we have already observed some black swans, we will not put a lot of trust in the hypothesis that all swans are white. But there is a sense of confirmation according to which the observation of another white swan lends some support to the claim that all swans are white, if we take that piece of evidence in isolation from the evidence we have previously gathered.

Richard Swinburne (1971) nicely shows the contrast between the two notions with the following example. Take the hypothesis: "All grasshoppers are located outside the county of Yorkshire." If I observe a grasshopper just over the border of the county, this observation should serve as confirming evidence of the hypothesis. But given what I know about grasshoppers (that they jump here and there without paying too much attention to county borders), my observation makes it more likely that other grasshoppers have entered the county, thereby undermining the initial hypothesis. One and the same instance can be confirming the hypothesis according to one sense of "confirmation" (based on the entailment relation between evidence and hypothesis) and disconfirming it according to the other sense of "confirmation" (if background beliefs are also taken into consideration).

Apart from being insensitive to the distinction between these two notions of confirmation, the hypothetical-deductive model endorsed by Hempel, according to Salmon (1990), has a number of weaknesses. It does not seem to allow for the possibility that alternative hypotheses can be confirmed by the same observation, and that some of these hypotheses can be more plausible than others given that observation. At least in its explicit formulation, Hempel's model does not take into account the initial plausibility of the hypotheses subject to evaluation. And finally, it struggles to deal with the case of statistical hypotheses whose confirming evidence cannot be deduced from them. These problems have been addressed by turning to probability theory, and in particular *Bayes's Theorem*. Can its application to confirmation in science deliver better results?

The starting point is to assume that what we are interested in when we study scientific confirmation is the way in which a piece of evidence lends support to a hypothesis given the available background evidence. Deductive accounts of confirmation don't deal very well with those scientific hypotheses which are confirmed by an observation statement that is not a direct consequence of them. An answer to the limitations that deductive accounts have is to address confirmation with the resources of probability theory. When philosophers think about probability, they can have one of the following conceptions in mind: either the probability of a theory is objective, or it is subjective. In the former case, the probability of a theory being true depends on how things are, and, in the latter, on the confidence agents have in the truth of the theory.

The subjective account can be used to develop an alternative account of confirmation to Hempel's. Bayes's Theorem, in particular, can tell us how our attitude to our current beliefs changes when new evidence becomes available (belief updating). The probability of a hypothesis *H* given new evidence *E* is equal to the probability of *E* given *H* times the probability of *H*, all divided by the probability of *E*.

Bayes's Theorem: $P(H/E) = [P(E/H) \, P(H)] / P(E)$

Here is an illustration of how you can apply the theorem to a concrete problem. Suppose Asja lives in a house with a garden, and her neighbor Tim's house also has a garden. In both gardens there are flowers of different types. If someone picks flowers randomly, the probability of picking a bluebell from Tim's garden is 1/2 and the probability of picking a bluebell from Asja's garden is 1/4. If John picked a flower from one of the two gardens at random and it is a bluebell, what is the probability that he picked it from Asja's garden?

P(A2) is the probability that John picked a flower from Asja's garden.

P(T1) is the probability that John picked a flower from Tim's garden.

P(B) is the probability that a bluebell is picked.

$$P(A2|B) = \frac{P(B|A2)P(A2)}{P(B|T1)P(T1) + P(B|A2)P(A2)}$$

Now, the probability of picking a bluebell in Asja's garden (P(B/A2)) is 1/4. The probability of picking a bluebell in Tim's garden (P(B/T1)) is 1/2.

$$P(A2|B) = \frac{\frac{1}{4} P(A2)}{\frac{1}{2} P(T1) + \frac{1}{4} P(A2)}$$

The probability of John picking flowers from Tim's garden (P(T1)) is 1/2 (as he chose randomly between the two gardens). And so is the probability of John picking flowers from Asja's garden (P(A2)).

$$P(A2|B) = \frac{\frac{1}{4}\frac{1}{2}}{\frac{1}{2}\frac{1}{2} + \frac{1}{4}\frac{1}{2}} = 1/3$$

So the probability that John picked the bluebell from Asja's garden is one third.

How does the application of this theorem help us understand scientific confirmation? By providing the resources to analyze the relation between some new information (a new piece of evidence) and a hypothesis. If a new piece of evidence is irrelevant to the hypothesis to be tested, then it neither confirms it nor disconfirms it. The evidence is neutral with respect to the hypothesis and the probability of the hypothesis given the (new) evidence (*posterior probability*) is equal to the probability of the hypothesis before the evidence became available (*prior probability*).

Neutrality (or Evidential Irrelevance): P(H/E) = P(H)

If a new piece of evidence lends support to the hypothesis to be tested, we can say that the *posterior probability* of the hypothesis will be greater than its *prior probability*. To go back to ravens, my degree of belief in the hypothesis that all ravens are black will be greater after I have observed another black raven. Here is another example from plate tectonics: the hypothesis of the continental drift was accepted long after it was first developed and its acceptance was due to newly acquired evidence about the nature of the geological mechanisms by which continents could shift across the surface of the earth. Thanks to the discovery of these geomagnetic anomalies (and other relevant pieces of evidence), the realization that continents could move as an effect of thermal convection made the theory of the continental drift more plausible. The probability of the hypothesis of the continental drift after the discovery of geomagnetic anomalies is greater than its probability prior to the discovery of those anomalies. The new evidence confirms the hypothesis.

Confirmation: $P(H/E) > P(H)$

The evidence undermines the hypothesis if the probability of the hypothesis before the evidence was available is greater than the probability of the hypothesis given the evidence. The observation of a white raven would make the probability of the hypothesis fall to zero, thereby not just disconfirming but falsifying the hypothesis that all ravens are black.

Here is another example of disconfirmation from atomic theory: at the beginning of the nineteenth century Dalton formulated the hypothesis that all matter is composed of small indivisible particles called atoms. When Thompson experimented with X-rays at the end of the nineteenth century, he found that atoms were not indivisible particles, and that they were in turn composed of smaller particles, e.g. electrons moving fast around a nucleus. Thompson's observation has made the probability of Dalton's hypothesis that atoms are small indivisible particles less probable, thereby disconfirming it.

Disconfirmation: $P(H/E) < P(H)$

Notice that, differently from Hempel's model of confirmation, the Bayesian approach can offer an indication of the *extent* to which a hypothesis is supported by a new piece of evidence. This seems to be an advantage. There are also other results of the application of Bayes's Theorem to scientific confirmation which are very promising. For instance, they can account for the following facts about confirmation:

a. Confirmation is greater when the probability of the evidence independent of the hypothesis is greater (e.g. *unexpected* results confirming the hypothesis confirm it to a higher degree).
b. If a hypothesis entails the evidence and counter-evidence is found, then the hypothesis is falsified because posterior probability is 0 ("All ravens are black" entails "This raven is black").
c. A universal hypothesis that all Fs are Gs is confirmed by the observation of a non-G non-F, but to a much less significant extent than the observation of an F which is G.

The latter point tells us that for Bayesians as well as for Hempel the paradox of the ravens is not really a paradox. Bayesians also conclude that we should accept the counter-intuitive claim that the observation of a white shoe is confirming evidence for the hypothesis that all ravens are black. As Patrick Maher (2004) puts it, the observation of a white shoe weakly confirms that all ravens are black because a white shoe is not a *counter-example* of the hypothesis that all ravens are black. What Bayesians claim they can do and Hempel's model could not do is *measure* the degree to which the observation confirms the hypothesis and maintain that a positive instance (a black raven) confirms the hypothesis that all ravens are black to a much greater extent than a contrapositive instance (a non-black non-raven).

Discuss: Do you think that the notions of confirmation and disconfirmation as formally defined by Bayesian probability map onto the way in which these notions are used in science?

Critics of the Bayesian approach to confirmation argue that there are other aspects of the relationship between hypotheses and evidence that this approach does not have the resources to adequately represent. They are concerned that scientists do not really reason in a way that can be formalized through probability calculus. The worry is that human subjects do not seem to be good statisticians and tend to be conservative when assessing the import of new evidence (Kahneman et al. 1982; El-Gamal and Grether 1995) and that, both in the personal reports and *post-hoc* reconstructions of how scientists arrive at the acceptance of the theories they defend on the basis of the available evidence, very rarely is there any mention of probability (Kelly and Glymour 2004).

Exercise: Do you think the way scientists reason when they assess the plausibility of their theories is relevant to the project of defining theory confirmation in science?

Here we shall consider only one objection to the Bayesian approach to confirmation, raised by Clark Glymour (1980). This objection is known in the literature as the "problem of old evidence." Glymour observes that Bayes's Theorem says nothing about those theories that are confirmed by evidence that is already known. His own example is Einstein's relativity theory (*H*) in 1915 being confirmed by the anomalies in the perihelion of Mercury (*E*) which had been known for over a century. Can old evidence be taken to confirm a hypothesis within the Bayesian approach? It would seem that it cannot, if the probability of the evidence is 1 and, therefore, the prior probability of the hypothesis equals the probability of the hypothesis given the evidence.

Howson and Urbach (2006) defend Bayes's Theorem by arguing that the background knowledge which determines the prior probability of *H* should not be taken to include *E*. So the probability of the hypothesis should be relativized to existent background beliefs, with the exclusion of the belief in the potentially confirming evidence. This step would be justified according to the authors, because the purpose of the exercise is to measure the impact of *E* on the probability of *H*. However, the move leaves many questions unanswered: Isn't it a limitation of the account if it cannot deliver the results we expect in the case of the confirmation of a theory given *all* current background beliefs? And how exactly would we go about excluding *E* from the body of background knowledge?

 Exercise: Can you think of other problems for the Bayesian approach to confirmation?

The message we can derive so far from the discussion of two of the most influential approaches to the theory of confirmation is that any attempt to formalize the way in which scientists operate when they assess some hypotheses on the basis of new or previously acquired evidence runs into problems. Either the attempts of formalization give rise to inconsistencies and paradoxes, or they fail to capture what scientists actually do, and see themselves as doing, when they test theories. Although the reconstructions offered by the hypothetical-deductive model and the Bayesian probabilistic approach provide useful insights into the notion of confirmation and its difficulties, they do not seem to be fully satisfactory accounts of the *practice* of scientific confirmation.

3.2.3.   The new riddle of induction

The problems Hempel faced with his account of theory confirmation were a concern for Nelson Goodman as well. Goodman (1954, 2006) presents a new riddle which emphasizes some difficulties with induction to a universal generalization. Can the observation of a black raven ever *confirm* the general statement that all ravens are black?

Goodman introduces a new term, "grue," that stands for the property of being green up to a certain time in the future (let's say, December 31, 2080) and then blue after that time. The statement "All the emeralds I have observed so far are green" seems to inductively support the hypothesis that "All emeralds are green," but it could as well inductively support the hypothesis that "All emeralds are *grue.*" This means that we should be wary about relying on induction to justify our scientific hypotheses, as in some cases the same evidential bases can lend support to two general hypotheses that generate competing predictions of future observations.

With Hempel as his polemical target, Goodman wants to show that it is not promising to search for an answer to the problem of induction and for an account of confirmation in science by examining the *syntactical* features of the statements reporting hypotheses and observations, and the logical relations between these statements. It is the *semantics* that matters: according to Goodman, only general statements of a certain kind, *lawlike* statements, can be supported by particular observations of their instances. We cannot establish whether a statement is lawlike just by glancing at its syntactical form. We need to pay closer attention to the semantic properties of the predicates it contains.

This is how the problem can be formulated. All emeralds observed before December 31, 2080 are green. We expect the next emerald to be observed after that date to be also green because we trust the general statement that all emeralds are green. All emeralds observed before December 31, 2080 are also grue, given the meaning of this predicate, but somehow we do not trust the prediction that all emeralds will be grue after that date. What makes it the case that two statements are equally confirmed by the observations conducted until the present time, but only one of them generates predictions that can be confidently projected into the future?

Table 3.2 Non-artificial predicates which have behaved a bit like "grue."

Predicate	History
species	Before the acceptance of Darwinism, "species" was meant to designate a group of organisms with some common features. Then theoretical biologists came to believe that boundaries between species are vague and vary across time. Today there is still a lively debate on whether "species" is a genuine natural kind term and how to best characterize the conditions something needs to satisfy in order to be a species. "Essentialism" about species cannot be defended, because members of the same species do not necessarily share intrinsic properties. For instance, Sober argues that species is a historical entity, given that two organisms belong to the same species in virtue of their historical connection.
jade	"Jade" has been used as a generic term for two gems with different chemical composition, nephrite and jadeite. Jadeite and nephrite were not distinguished until the beginning of the 19th century because they have very similar appearance and share other superficial properties (e.g. toughness). Now they are distinguished by mineralogists because they are made up of different silicate materials and have different characteristics (e.g. jadeite is much rarer than, and comes in different colors from, nephrite).

3.2.4. Attempted solutions to Goodman's riddle

Is there something wrong with "grue"?
Philosophers thinking about theory confirmation and induction have worried that the riddle put forward by Goodman is determined by the introduction of a predicate which is artificially created, and that the considerations which apply to that predicate do not apply to the predicates we use in everyday language. The predicate "grue" seems artificial, because it is disjunctive (that is, it contains an "either. . .or. . ." clause).

But Goodman observes that the impression of artificiality is due to our habit of regarding the predicate "green" as primitive and the predicate "grue" as derivative. If we abandon this assumption, and take "grue" and "bleen" as primitive (where "bleen" means "blue until December 31, 2080 and green afterwards"), then we can define "green" as "grue before December 31, 2080 and bleen afterwards." The so-defined predicate "green" would be disjunctive, and, by the line of reasoning above, more artificial than "grue".

If we pay attention to the shifts in meaning of predicates which are used in everyday language and science, we can find that some of these predicates behave a bit like "grue." What the terms mean and what they refer to has changed in time together with the theoretical descriptions associated with them. This, in turn, has had an effect on the perceived projectibility of the generalizations containing these terms.

 Exercise: Can you think of another example of a general statement that is true now but cannot be confidently projected into the future because it contains a grue-like predicate?

Here is another related line of argument: the predicate "grue" is artificial not because it is disjunctive, but because it is a made-up predicate and it is not embedded in our language. Goodman defines a predicate as projectible if we can trust our inductive inferences containing that predicate, because the predicate applies now to the same objects to which it was found to apply in the past and we have confidence that it will continue to do so in the future. A predicate cannot be confidently projected, according to Goodman, unless it is entrenched in our everyday language and is commonly used. The problem with "grue" is that it lacks entrenchment.

But, as Colin Howson (2000) observes, some of the predicates that are used in hypotheses which receive excellent confirmation are brand new terms – he refers to the case of scandium, a rare earth metal which was so named in 1879. The fact that hypotheses about scandium are well confirmed renders speakers confident in projecting them onto the future, whether or not the first formulations of those hypotheses contained unfamiliar terms. Even if terms like "scandium" don't have an established track-record when they are first used and they have not originated in everyday language, they are projectible according to Goodman's definition, which suggests that entrenchment is not a necessary condition for projectibility.

Exercise: Find other examples of newly introduced theoretical terms that appear in scientific generalizations.

Lawlikeness
The difference between lawlike statements and accidental generalizations might explain why induction does not seem to work with predicates such as "grue." The statement "All emeralds are grue" is not a lawlike statement but an accidental generalization. If a generalization is accidental, it cannot receive confirmation from the observation of one of its instances.

Think about a particular case. "This man in the blue shirt is unmarried" does not confirm the following generalization: "All men in blue shirts are unmarried." One of the reasons why we might think that the particular statement does not support the general statement is that there is no special relation between the predicates in the statements. What one wears seems to have no special connection to one's marital status unless there is an explicit convention in place which establishes such a connection – e.g.

sometimes we can tell whether someone is a Catholic priest or a nun by observing what they wear, and we can make inferences from their roles to their marital status. But in our original example, it is *accidental* that the man we met at the party was in a blue shirt and was unmarried.

Compare with the statement "This ruby is red" which seems to confirm the hypothesis "All rubies are red." There seems to be a special relation between being a ruby and being red. Rubies, as many gemstones, get their color due to the impurities contained in their framework (e.g. rubies are crystals of corundum which contain impurities of chromium). So it is *not* accidental that the ruby I have just observed is red.

Now, let's return to "grue." Is there a principled reason to regard "All emeralds are green" as a lawlike statement and "All emeralds are grue" as an accidental generalization? Usually lawlike statements are distinguished from accidental generalizations, because the former don't make any explicit reference to a particular object, place or time. "All strawberries in my fridge are ripe" would count as an accidental generalization, because it contains a clause that specifies the location of the strawberries to which I attribute the predicate of being ripe. As we saw before, the definition of "grue" contains some reference to time, but so does the definition of "green" if we define it on the basis of "grue" and "bleen." This criterion (which raises independent problems) does not seem to help us solve the riddle.

 Exercise: Think of other limitations of this first attempt to tell accidental generalizations apart from lawlike statements.

 Discuss: Can there be laws that contain spatio-temporal qualifications?

Suppose we do find a reliable and principled way to distinguish accidental generalizations from lawlike statements. It could still be objected to this attempted solution that, when the right background beliefs are in place, general statements receive confirmation on the basis of their observed instances independently of there being a special relation between the predicates contained in the generalizations. The thought is that lawlikeness is sufficient but not necessary for projectibility. Even if no assumption about the nature of the relation between being a ruby and being red could be made before we knew about what determines color in gemstones, the generalization "All rubies are red" was already being safely projected into the future.

It seems that we still lack a good reason to believe that "All emeralds are grue" is not going to generate accurate predictions about future observations of emeralds.

3.3. Models of Explanation

In this section we shall review some philosophical approaches to the notion of scientific explanation. According to the *nomological model*, an event can be explained only if it is subsumed under a law or a statistical generalization and explanation has the form of a deductive or inductive argument. In the *causal model*, scientific explanation is about identifying causal relations, causal chains, or common causes between events. The *pragmatic account* is more interested in the way in which the grammar of explanations work than in offering a unified characterization of all instances of scientific explanation.

3.3.1. Hempel's models of explanation

The Deductive-Nomological Model of Explanation (DN) is based on the idea that any explanation is constituted by:

- an *explanandum* (something that needs to be explained)
- an *explanans* (something that does the explaining).

The *explanans* must be true and the *explanandum* must be one of its logical consequences for the explanation to succeed. The model is called "deductive" because the explanation has the form of a deductive argument; and it is called "nomological" because the *explanans* must contain at least a law of nature ("nomos" = norm).

A singular event E is explained if and only if a description of E is the conclusion of a valid deductive argument, whose premises involve a lawlike statement and a set of initial conditions. The model is also referred to as the "covering law" model because the occurrence of the event to be explained needs to fall under the scope of a law of nature. The idea is that you cannot explain a particular event unless you can deduce it from a lawlike generalization – accidental generalizations won't do. But as we saw with respect to one of the attempts to solve Goodman's riddle, it is extremely challenging to find a principled way to distinguish accidental generalizations from laws, and this can be regarded as a weakness of the DN model.

Here is an (extremely simplified) example of the application of the model to the explanation of a particular phenomenon.

> *Initial Conditions*: (i) Tectonic plates A and B rubbed together at time T_1 at location P_1.
> *Lawlike generalizations*: (a) When tectonic plates rub together, the movement transmits waves of energy to the surface of the earth. (b) When energy is transmitted to the surface of the earth, tremors and shakes occur. (c) etc.
> *Explanandum*: There was an earthquake at time T_2 at location P_2.

Four conditions of adequacy apply to explanations of this sort for Hempel.

1. *The argument must be valid.* The occurrence of the earthquake at a particular time and in a particular place is *entailed* by the initial conditions and the lawlike generalization. If the premises of the arguments are true, so is the conclusion.

2. *The premises must include a lawlike statement.* Some of the premises of the argument above need to be lawlike generalizations: e.g. "When energy is transmitted to the surface of the earth, tremors and shakes occur."
3. *The premises must have empirical content and be verifiable.* Premises have empirical content and can be verified, although there might be issues emerging from the attempt to verify (rather than confirm) a lawlike generalization when it is expressed by a universal statement.
4. *The premises must be true.* This is a different condition from (1)–(3). If it is not met, then the argument from the explanans to the explanandum still has the right form, that is, it is a *potential* explanation. If condition (4) is satisfied, then the argument is no longer just valid, but it is also sound, and it is an *actual* explanation.

Hempel claims that the model can account for causal explanation, and that it reveals the symmetry between explanation and prediction. Let's see these two points in turn. Hempel argues that causal explanation is just a type of explanation that can be represented via the DN model. As he did with the relation of confirmation between an observation statement and a hypothesis, Hempel accounts for the relation of cause and effect via the logical relation of entailment. Just as an observation statement follows from the hypothesis that it confirms, so an event to be explained follows from the initial conditions and the lawlike statements that contribute to its explanation. The symmetry between explanation and prediction is also a result of this approach. The lawlike statement and the initial conditions explain the event which is described in the conclusion of the argument, but they also predict it. In the example above, in a backward-looking exercise, we have explained the occurrence of the earthquake on the basis of the consideration of the initial conditions and the application of the universal generalization. But similarly, in a forward-looking exercise, we could predict the occurrence of the earthquake on the basis of the initial condition and the lawlike generalization.

A similar argumentative structure, according to Hempel, applies also to statistical hypotheses, but the difference in that case is that the generalization makes the conclusion not entailed by the premises, but more probable given the premises (this is the Deductive-Statistical Model). DS is appropriate when we want to explain a statistical hypothesis and we can do it deductively if one of the premises is a statistical generalization. When the event to be explained is a one-off case, then the deductive model cannot be used and we have to appeal to the Inductive-Statistical Model of Explanation. The conclusion is regarded as highly probable given the premises but does not follow from the premises. In the DN model one condition of adequacy for the explanation was that the *explanans* had to contain a lawlike statement. For the IS model the condition is that the generalization contained in the

Table 3.3 Inductive Statistical Model of Explanation.

Structure of IS	Example of IS
Fa	Sunita takes antibiotics to treat an infection.
P(G/F) is very high	The probability of recovering from an infection by taking antibiotics is very high.
Then: G(a)	Then it is very likely that Sunita will recover.

Table 3.4 Other examples of explanation.

Structure of explanation		Example 1	Example 2
Premise 1	*Initial conditions*	Height of the flagpole; elevation the Sun; etc.	Expansionism of Germany; tension between Austro-Hungarians and Serbians; etc.
Premise 2	*Lawlike generalizations or inductive generalizations*	Set of laws of optics (e.g. Light travels in a straight line) plus laws of geometry	Generalizations (e.g. nations participate in international conflicts to defend their commercial interests)
Conclusion	*Event to be explained*	Length of the shadow cast by the flagpole	Occurrence of World War I

explanans must have a very high probability. The IS argument can also be used for the purposes of prediction: to say that the premises explain the event is also to say that we expect the event to happen given the premises.

 Exercise: Identify other explanations that fit either the DN model or the IS model.

Hempel's models of explanation have been criticized for failing to specify plausible necessary and sufficient conditions for satisfactory scientific explanations. According to the critics, the reason why some DN or IS arguments are not truly explanatory is that they do not point at the genuine cause of the event to be explained. In other cases, we have perfectly satisfactory explanations that do not fit the DN model (e.g. because the event to be explained does not follow under the scope of a lawlike generalization) or the IS model (e.g. because the generalization that would explain the event has low probability). We shall examine two objections in more detail: relevance and symmetry.

3.3.2. *Relevance, symmetry, and causal relations*

Wesley Salmon (1989) raised the problem of *relevance* with the following example:

> *Initial condition*: Butch takes birth control pills.
> *Generalization*: People who take birth control pills do not become pregnant.
> *Explanandum*: Butch has not become pregnant.

The argument above qualifies as explanatory on Hempel's model, but the first premise seems to be irrelevant to the conclusion. The reason why Butch has not become pregnant has nothing to do with whether he took birth control pills, but it has to do with his being male.

Consider the following example which illustrates the same issue:

> *Initial condition*: Emma takes vitamin C for a week to treat a cold.
> *Generalization*: The probability of recovering from a cold after taking vitamin C for a week is very high.
> *Explanandum*: Then it is very likely that Emma will recover.

Even if the generalization is highly probable, the explanation is a bad one. Colds usually clear in a week anyway, with or without regular intake of vitamin C. The high probability of the generalization was introduced by Hempel as a condition for successful IS arguments in order to guarantee that there was good evidential basis for the inductive inference. But the high probability of the second premise does not actually guarantee that the explanation is a good one. Some additional criterion of relevance needs to be in place, because the satisfaction of the formal requirements for an adequate explanation is not sufficient. Some philosophers of science (see Psillos 2002) argue that a reference to causal relations could help us discriminate between relevant and irrelevant explanations, but that this type of relations is not captured by Hempel's model.

Sylvain Bromberger (1966) raised the problem of *symmetry* with the DN model of scientific explanation. Here is a classic example:

> *Initial Condition*. The barometer is falling rapidly.
> *Generalization*. Whenever the barometer falls rapidly, a storm is approaching.
> *Explanandum*. A storm is approaching.

The falling barometer is an indicator of an approaching storm, but it does not explain the occurrence of the storm. It is more intuitive to say that it is the approaching storm that explains the falling barometer. This case breaks up the desired symmetry between explanation and prediction. By using the barometer we can *predict that* there will be a storm, but we don't *explain why* it occurs, as both the barometer dropping and the occurrence of the storm are due to something else, a change in pressure. Hempel cannot account for the asymmetry between cause and effect within deductive-nomological explanations.

The same problem emerges with the flagpole example found in table

3.4. We can explain the length of the shadow by reference to optical laws, geometry, and the height of the flagpole. Notice that we can also "explain" the height of the flagpole by reference to the length of its shadow, and this direction of explanation seems counter intuitive:

> *Initial Condition*: Length of shadow.
> *Laws*: Laws of optics and geometry.
> *Explanandum*: Height of flagpole.

These examples suggest that Hempel's model does not have the resources to discriminate the roles of initial conditions and *explanandum*, as each of them can be used to derive the other in conjunction with the laws or generalizations used as the second premise of the argument. The analysis of this example makes explicit an important disanalogy between the relation between *initial conditions* and *explanandum* in commonsense (which is not perceived as symmetrical) and the relation between *initial conditions* and *explanandum* in the deductive model (which is perfectly symmetrical).

Some attempts have been made to solve the relevance and the symmetry problems by specifying further conditions that the relation between *explanans* and *explanandum* needs to satisfy. The causal model of explanation developed by Salmon and others can overcome some of the problems found in Hempel's model of explanation, such as symmetry, as it is based on the idea that the relation between *explanans* and *explanandum* is a causal relation. Salmon's idea is that all events are part of causal chains and are related to each other via spatio-temporal continuity and statistical relevance. When values correlate, we assume that they have a prior common cause. For instance, the pressure falling can explain both the occurrence of the storm and the dropping of the barometer. But maybe other events, prior to the change in pressure, are the common cause of such a change and the behavior of the barometer. This method for establishing causes might bring us to a regress of explanation: when do we stop searching for a common cause?

The main difficulty that the causal account encounters is the clarification of the notion of causing. To say that *A* caused *B* is not to say that *A* is necessary for the occurrence of *B*, as *B* might have occurred independently of *A*. The following example illustrates this case. A man ingests arsenic and dies 24 hours later. The poisoning is the cause of his death, but he might have died nonetheless if he had forgotten to pay attention when crossing the road, or if he had been struck by lightning.

A counterfactual account of causation suggests that whenever "*A* caused *B*" is true, it is also true that "If *A* had not happened, then *B* wouldn't have happened." But this account has other problems, as antecedent conditions to *A* might be in the same counterfactual relation to *B* and yet not be relevant to a causal explanation of why *B* occurred. If it is true that ingesting arsenic caused the man's death, then the death would not have occurred unless the man had been poisoned. But it is also true that the death would not have occurred if the man had not met his brother (who poisoned him) and yet meeting his brother does not seem to be *the* cause of his death. We need a way of focusing on the salient factors in the causal chain of events that lead to the man's death if we want to provide a satisfactory explanation of it.

 Exercise: Can you think of good scientific explanations that do not mention causes?

 Discuss: Do causal models fare better than Hempel's model in providing necessary and sufficient conditions for scientific explanation?

3.3.3. *A pragmatic approach to explanation*

Many of the problems raised by Hempel's theory of explanation, including the two we have presented (relevance and symmetry), can be at least partially addressed by reference to a causal relation between the *explanans* and the *explanandum*. But some theories seem to explain without providing information about causal processes (e.g. elementary quantum mechanics; cognitive psychology; geometry) and this raises concerns about whether scientific explanations *need* to be causal. Should causal mechanisms have a privileged role in explanation?

Bas Van Fraassen (1980) reviews the literature on explanation, from Hempel's models to the causal accounts, and challenges the assumption made by both nomological and causal accounts of explanation, that explanation is a relation between the event to be explained and a hypothesis. He develops a pragmatic model of explanation, according to which a fact explains another fact relative to a theory which is accepted. Important assumptions in this account are that (1) the theory used to provide an explanation of the facts we are interested in does not need to be true, or even empirically adequate, in order to play the required explanatory role; and (2) explanations are always relative to a context, which means that it would be wrong to suppose that, for any fact that demands an explanation, there is just one satisfactory answer to the question(s): "How (why) did *F* happen?"

This analysis points at two sources of context-dependence. First, an explanation is always relative to a *theory* and the *interests* of the people looking for an explanation. These two elements characterize the context within which we can assess the salience and relevance of alternative explanatory factors. Second, explanation is relative to which events we consider to be a relevant alternative to the event we need to explain.

Let's imagine a murder investigation. The question: "What caused the death of this person?" will have different answers for the medical doctor conducting the *post-mortem* examination and the detective searching for a murderer with motive. This determines the explanatory relevance of the hypotheses that will be put forward. In the former context, alternative explanatory hypotheses for the death of the murder victim might include, for instance, drowning and being shot. In the latter context, the detective is searching for plausible motives for homicide, and alternative explanatory hypotheses might include, for instance, the hope to inherit a fortune or jealousy.

Van Fraassen presents another example in which a why-question can be answered in different ways, depending on the interest of the person asking the question. We could answer the question "Why does blood circulate through the body?" either by reference to the function of blood circulation ("*In order to* bring oxygen to body tissues") or by reference to the mechanism that makes the blood circulation possible ("*Because* the heart pumps the blood through the arteries").

When we reflect about the event to be explained, we contrast it with other events that might have happened instead, and ask why the original event (and not one of the alternatives we have considered) did actually occur. The detective might look for an explanation why Mr Brown was murdered in his house, rather than in his office, or why he was murdered at 8pm rather than at 6am. The anatomist might wonder why the blood circulates at speed x rather than at speed y. The evolutionary biologist might think about why blood circulation developed instead of other ways there could have been to guarantee supply of oxygen to the body tissues. Possible events that differ from the actual event to be explained in one respect or other (e.g. the location of the crime; the time of the murder) constitute the relevant *contrast class*.

What makes an explanation good? There are three main criteria. First, an answer to a why-question such as "*E* occurred because of *A*" can be evaluated in its own right on the basis of *truth or plausibility*. In the murder case, an explanation for the death of Mr Brown is that Miss Smith shot him. The hypothesis that Miss Smith shot the victim might lose plausibility if we discover that Miss Smith has never owned or used a gun in her life.

Second, the answer can be evaluated with respect to its *relevance* to the event to be explained. Does *A* really favor *E* with respect to its contrast class (those events that could have occurred instead of *E*)? If the victim had been Miss Miles, who is Miss Smith's aunt and would have left her a large inheritance, then Miss Smith's motive for murder would have been clear and the explanation more convincing. But the victim was Mr Brown who had no known relationship with Miss Smith.

Third, the answer can be evaluated with respect to other possible answers to the why-question – e.g. is "Because *A*" more probable than "Because *B*" given the information available to us? If another suspect with a strong motive and no alibi, Mr Jones, is found to be a skilled gunman, then the probability of Miss Smith having committed the crime does not seem to be as high as the probability of Mr Jones having committed the crime.

In the pragmatic account, explanation of an event *E* is de-mystified: it is seen just as an answer to a why-question which is satisfying relative to a background theory which determines the range of alternative hypotheses and the respects in which the event needs an explanation. According to Van Fraassen there seems to be no principled way to distinguish explanation in science from other types of explanation, if not by reference to the type of events that demand explanation and the range of background theories on the basis of which the relevant explanatory hypothesis is selected.

 Exercise: What makes Van Fraassen's account of explanation pragmatic?

Summary
In this chapter we have seen models that have been designed to account for the distinctive way in which knowledge is acquired, consolidated and put to use in science. The attention that the logical positivists pay to the logical structure and syntactical features of confirmation and explanation is supposed to be part of a general project to account for the practice of science in a purely objective way (independent of sociological and psychological facts about discovery or justification) and to explicate the concepts used by scientists in a clear and unambiguous fashion.

But, even if their analyses are very illuminating, and help us draw some general conclusions about the way in which scientists operate, the attempt to explicate confirmation and explanation in terms of mere logical relations between statements fails to capture all the features of confirmation and explanation that are significant in the practice of science.

Goodman and Van Fraassen argue that semantic and pragmatic considerations need to be taken into account in order to make sense of the way in which theories are formed, tested, accepted, and applied in science. This does not necessarily speak against the objectivity of science, and it is not a threat to viewing science as progressive and directed towards truth. But it might lead us to accept that there are different notions of confirmation and different types of explanation that serve different and sometimes equally valuable purposes.

The pragmatic approach generates further doubts about the legitimacy of a well-defined demarcation criterion between science and non-science. If it is true at least some of the time that theories are developed via models whose role is to prompt analysis and reflection, and to test hypotheses, and that the adequacy of an explanation should be conceived as relative to a context, and hostage to our expectations, then the perceived gulf between the practice of natural sciences and human or social sciences, and the practice of science in general and everyday reasoning, seems to be thinning down to the point that the elements of continuity outweigh the marks of distinction.

Preview of future attractions
The debate on theory and observation will continue in the next chapter where we shall ask whether scientific theories describe and represent

reality as it is, or just provide useful tools for the prediction and manipulation of nature.

The difficulties in developing a satisfactory formal account of theory confirmation and explanation will inform our discussion of the rationality of scientific change in chapter 5. Can we really choose between rival theories based on their empirical adequacy and their explanatory power, or are there other factors that determine theory choice?

Issues to think about

1. What is the connection between Goodman's riddle and Hume's problem of induction?
2. Is "species" a predicate that can be projected?
3. What model of explanation is better suited to the social sciences?
4. What makes a model of explanation "nomological"?
5. Does Bayes's Theorem offer a reconstruction of how we *do* update our beliefs, or how we *would* update them, if we were rational agents?
6. Is there a necessary connection between being a ruby and being red?

Further resources

If you want to explore the literature on the nature of scientific theories, a good place to start is the entry "Theories" by Giere in the Blackwell *Companion to the Philosophy of Science* (2000). Also, there is an excellent entry on "Models in Science" by Frigg and Hartmann in the *Stanford Encyclopedia of Philosophy* (2006). Recommended texts on the syntactical view of theories are Carnap (1966, chapters 23–6) and Hempel (1970); on the semantic view, see Suppe (1989) and Van Fraassen (1980, chapter 3).

If you want to know more about attempted solutions to the paradoxes of confirmation, start with the survey by Swinburne (1971) and the collection on "grue" edited by Stalker (1994). You can also tackle the classics by Hempel (1945) and Goodman (1954, 2006). A fourth edition of the classic text by Goodman, *Fact, Fiction and Forecast*, is available with a foreword by Hilary Putnam. See Parts III and IV for, respectively, a statement of the grue paradox and an attempted solution to it.

If you are new to probability and want to gain a deeper understanding of probabilistic approaches to confirmation, Hacking (2001) will be useful. For a detailed discussion of the probabilistic accounts of confirmation and proofs of some of the claims presented in the section on Bayesianism, look up both Maher (2004) and Howson and Urbach (2006).

Critical overviews of the literature on models of explanation can be found in Salmon (1989) and Van Fraassen (1980). To learn about computational models of explanation, see Thagard and Litt (2008).

4 Language and Reality

Here we are going to explore some of the issues raised by the use of language in scientific practice and scientific theorizing. Some of these issues won't be entirely new, and they will follow nicely from our previous discussion of the nature of scientific theories. Is there a coherent and meaningful distinction between theoretical and observational terms? How does the meaning of terms like "book" (which denotes an artifact) differ from the meaning of terms like "oxygen" or "marriage" (which denote respectively a natural and a social kind)?

In order to understand the motivation for introducing these distinctions, we need to know a bit more about how language in general works. There are conflicting theories about the way in which the terms we use get their reference and how something in our minds, an idea or a concept, gets to refer to or pick out an object in the world. The debate between causal theorists of reference and descriptivists will be introduced briefly and it will set the scene for the discussion of more specific issues in the philosophy of science. Can competing theories be compared if the theoretical terms they use refer to different entities? Are all the entities posited by our currently accepted theories really *out there*?

How reference works matters to the enterprise of comparing scientific theories, because when theories are overthrown and replaced, some of the terms deployed in the overthrown theory are maintained but can be associated with different theoretical descriptions. Other terms lose their referents entirely: in the new theory there might be no room for some of the entities whose existence was accepted before. In any case, theory change has a significant impact both on the language and the ontology used by scientists and laypeople in a community.

When the change due to overthrowing a previously accepted theory is fairly radical, and new entities are posited for the purposes of explanation and prediction, or the entities posited by the old theory are assigned significantly different theoretical descriptions, there can be serious concerns about the effectiveness of communication among scientists committed to competing theories. Can we translate statements of a theory into statements of another theory and preserve genuine communication? If the answer is negative, the very possibility of comparing the competing theories (and thereby assessing them on the basis of criteria such as explanatory power and simplicity) is undermined.

Questions about the meaning and reference of theoretical terms and about the possibility of comparing theories re-emerge in the debate on scientific realism. What is the relation between science and the world in which we live? Are scientific theories, or at least the scientific theories that we currently accept, supposed to offer a description of how the world really is? If so, can we say that scientific theories offer a better, maybe more fundamental, description and representation of reality than alternative means of description and representation? We shall view scientific realism in the context of the wider debate on realism in philosophy and then present a range of positions: there are full-blown realist positions based on the arguments from the success of science; there are instrumentalist, relativist, or constructivist positions that put pressure on the assumption that science is a guide to reality; and finally there are influential middle-ground positions which concede that full-blown realism is unsupported but resist the radical consequences of the anti-realist alternatives.

By the end of this chapter you will be able to:

- Revisit the relation between observation and theory by reference to the language of scientific theories.
- Explain how the language of science changes when the accepted theoretical descriptions change.
- Distinguish two influential theories of meaning and reference.
- Account for the distinction between natural-kind terms and non-natural-kind terms.
- Assess the hypothesis that competing dominant theories are incommensurable.
- Discuss the role of scientific theories and how they relate to reality.
- Assess different arguments for realism and anti-realism.
- Come to an informed opinion about the ontological status of theoretical terms.

4.1.   Meaning, Reference, and Natural Kinds

Here we shall introduce the notions of meaning and reference and explore the view that natural-kind terms obtain their reference in a way that depends on the essential properties of the natural kind that they name.

4.1.1.   How terms get their meaning

When we think about the meaning of a word, many things spring to mind. The word "carrot" names a vegetable of a certain color, shape, and size that humans and other animals sometimes eat. If asked to explain what "carrot" means to someone who is learning English, we can draw a carrot on a piece of paper as an illustration, show them a carrot or list properties that typically carrots have, e.g. they are orange, they can be cooked or eaten raw, they contain vitamins, they are recommended to enhance natural tanning etc. Note that the latter option is viable if the person already has some competence in the English language. Other types of competence might be relevant: users of the term "carrot" know something about carrots, but people growing carrots in their vegetable gardens, nutrition experts, or writers of cookery books know much more about carrots than other people do, and would be able to list properties of carrots that are not regarded as stereotypical of carrots.

Philosophers of language emphasize two aspects of meaning: (1) the fact that (at least some) words refer to specific objects in the world (reference or extension) and (2) the fact that speakers associate a description to those words (sense or intension). Available theories of meaning account for these two facts, but two theories fight over what determines reference for proper names, names such as "Max," "Paris," and "Neptune." For *descriptivism*, the description that the speaker associates with the word determines the reference of the name. For the rival theory, the *causal theory of reference*, what determines the reference of the name is some non-mental facts about the way in which the name relates to the object it refers to.

The descriptivist theory says that the reference of a proper name is determined by the description (or set of descriptions) that speakers associate with the name (Strawson 1957; Searle 1969). There are different versions of this theory. One view is that there is one privileged description that plays this reference-fixing role. The alternative is to suggest that there are several descriptions associated with the name and not all of them need to play the reference-fixing role or to be satisfied by an object for the name to successfully refer to that object ("cluster theory"). Descriptivist theories also differ in terms of whether the description is what an individual speaker associates with the name, or a relevantly defined group of speakers (e.g. a linguistic community). The name "Daniel Kahneman" refers to the person who satisfies the description speakers associate with the name. For those who don't know Daniel Kahneman personally but know of his work, he is the psychologist who studied the limitations of human reasoning and won a Nobel Prize for Economics in 2002. It is likely that, for his family members and neighbors, other descriptions will be more salient.

The competitor theory of reference, the causal theory, says that a name or a term refers to whatever has the right causal connection with it (Kripke 1980). "Daniel Kahneman" will refer to the person who was named "Daniel Kahneman" and who has been called by that name since then, on the basis of that initial act of *baptism*.

Exercise: Can you anticipate some of the problems that the two theories of reference sketched above might face? Which one of the two theories is more convincing?

Discuss: How can one determine which descriptions are more or less central or representative?

One problem that is much discussed in the literature, and is often seen as an objection to a purely descriptivist account of the reference of proper names, is that of inaccurate descriptions. Suppose that, for many years, speakers believed that "Homer" referred to the author of the Iliad. And then they discovered that there is no unique person who wrote the Iliad. What should they think about the reference of "Homer"? If the description they used to pick out the individual named "Homer" were inaccurate, then "Homer" does not refer. But if they take the name to refer to whoever was baptized as "Homer" and later identified with that name, then it does not matter that the description commonly associated with "Homer" was inaccurate, and the name does not necessarily lose its referent when the truth about the authorship of the Iliad is revealed. In this latter scenario, all speakers need to do is revise the description, and this has no immediate consequences for the determination of the referent of "Homer" or the question whether Homer existed.

In the case in which *all* that speakers know about Homer is that they thought that he had written the Iliad, and no other description is associated with the use of the name, the conclusion that "Homer" does refer to someone about whom they know nothing, and that has no (special) connection with the Iliad, seems unsatisfactory because it does not account for the way in which the name has been used before it became known that Homer was not the single author of the Iliad.

Some philosophers, often moved by cases such as these, suggest that we can develop an intermediate position between the causal and the descriptivist theories of reference. Gareth Evans (1973), for instance, believes both that the causal connections between the name and its use is important, and that some descriptions are central to our understanding of how the name has been used. He puts forward the thought that the referent of a proper name is what caused speakers to associate with it the prevalent description that governs the use of the name.

4.1.2.   Twin-Earth

According to theories of meaning inspired by descriptivism (Frege 1892; Russell 1905; Kuhn 1962, 1970), the description we associate with terms that designate fundamental kinds in nature determine what we think about when we use those terms. Natural kinds are often distinguished from collections of abstract objects (such as numbers) or artifacts (such as chairs), but also from social kinds (such as gender). The term "electron" is an example of a natural-kind term and, for the descriptivist, it refers to whatever entity its theoretical description picks out. If, for the layperson, electron is the negatively charged part of the atom, for a chemist or physicist the description used to identify the entities they think about when they talk about electrons will be richer and more detailed, because they have a better understanding of the properties of electrons.

For causal theories, the treatment of proper names is extended to natural-kind terms (Putnam 1975). The reference of "gold," "tiger," or "water" is given by what gold, tigers, and water are, not by what we know about them. The referent of "gold" will be the thing that was first baptized "gold" and then identified with that term, independent of the changes that its theoretical description underwent from the first time that people came into contact with gold to the current use of the term.

Exercise: Can the explanation of the baptism of natural kinds work also for entities that are not directly observable, such as electrons? Consider a theoretical term of your choice (for instance, "traceability" in economics or "quantum" in physics). What determines the referent of that term? In what conditions would the term change its referent?

Putnam (1975) devises a thought experiment (now extremely influential both in the literature on natural kinds and on mental content) to show that our intuitions about meaning and reference support a causal theory of reference for natural-kind terms. He invites us to imagine that there is another planet which is almost indistinguishable from planet Earth, Twin-Earth. The only difference between Earth and Twin-Earth is that the substance that we call "water" on Earth is H_2O and the substance that they call "water" on Twin-Earth, is XYZ (where this stands for its chemical composition). What is interesting about this imaginary case is that on both planets the superficial properties of the substance are the same: "water" denotes a liquid that has no color, has no odor, is thirst-quenching, can be found in rivers and so on.

Here we have a case in which the chemically unsophisticated inhabitants of Earth and Twin-Earth share the same description associated with the term "water" (e.g. what one finds in lakes and rivers) but the term picks out two different substances. On Earth, it refers to H_2O and on Twin-Earth to XYZ. If this is the correct way to interpret the thought experiment, it shows that descriptivism must be wrong, and intension (or sense) cannot determine extension (or reference), because one can have a term with the same intension which has a different extension depending on the environment in which it is used.

Putnam creates this scenario to get us to share the intuition that meaning depends both on the external world (e.g. what water is) and on the way in which linguistic labor is distributed in the community. Even if the lay person cannot distinguish water from the substance that runs in rivers on Twin-Earth, for natural-kind terms such as "water," the way in which we identify instances of water hangs on the capacity to identify its hidden structure, its essential properties, and not just its superficial characteristics such as color and odor. In other words, Putnam argues that meaning is not something private in an individual speaker's head but something determined by the nature of things in the world that surrounds us. That (often not apparent) nature is something we learn by doing science and determines the criteria experts use to establish whether something is water or something else.

Does it make sense to say that "water" has two extensions, H_2O on Earth and XYZ on Twin-Earth? Not for Putnam, as it would not make sense to say that there are two entities to which the name "Homer" refers the person who was believed to have written the Iliad single-handedly, and the group of people who contributed to its being written. Water is H_2O and the term "water" has always referred to H_2O on Earth, even when people could not recognize water on the basis of its chemical composition. In Kripke's terminology, "water" is a rigid designator, that is, its referent remains the same in all possible worlds and picks up what is essential about water, its hidden structure. Even in 1750 on Earth, when nobody knew the chemical composition of water, the term "water" did refer to H_2O, because it behaved like an indexical. "Water" was assigned to some particular uniform substance by the initial baptism that fixed the reference of the term: when someone said for the first time "*This* is water." As any indexical expression, its reference is determined by the world, by the features of what speakers pointed at when they said "This is water."

Discuss: (1) Is your intuition that before the discovery of oxygen and of the chemical composition of water, the term "water" referred to two different substances, on Earth and Twin-Earth? (2) Is your intuition that at that time the inhabitants of Earth and Twin-Earth meant different things by "water"?

4.1.3.   Intuitions about natural kinds

Descriptivists claim that they have different intuitions about the Twin-Earth case and defend a view according to which the referent of "water" is whatever substance satisfies an accepted theoretical description (or operational definition) of water. The determination of reference is relative to the theory that is currently adopted. It makes no sense to say that before 1750 speakers were referring to H_2O by using the term "water" because at that time no such a description could have been associated with water and nothing to that effect appeared in the best available theory of what water was.

Another challenge might be levied to the distinction between superficial and essential properties for natural kinds, which matters to reference fixing, and to the idea, implicit in the view defended by Putnam, that referents of natural-kind terms are determined by the *essential* rather than by the *superficial* properties of the natural kind. Which properties are essential and which are superficial is not easily established independently of a scientific theory which explains the phenomena involving a natural kind. Even if there were essential properties that could be identified on the basis of empirical investigation, they could be other than the hidden structure. For instance, that the hidden structure is essential to a natural kind makes sense for some physical or chemical kinds, but is implausible in the case of the kinds studied by the life sciences.

Discuss: Should we believe that some properties of water are more important than others and that they are essential to what water is? Should the fact that water is H_2O play a more important role in reference fixing than the fact that water can be drunk?

Reflections on the use of language and the way meanings are assigned lead to reflections on the way in which we identify the nature of those objects to which terms in our language refer. There is a lively debate in metaphysics about the way in which nature is carved up: do scientists *stipulate* or *discover* that water is H_2O? It is plausible that, when we realize via empirical investigation what the chemical composition of water is, we learn what water was all along. As Putnam and other essentialists about natural kinds believe, the acquired knowledge of the identity between water and H_2O is used to correct past uses of the term "water." We have discovered the *essence* of water.

However, in other cases of baptism, it does seem likely that scientists were not compelled to characterize natural kinds the way they did by a consideration of the essential properties of a kind, but that they had a choice. The term "jade" refers to jadeite and nephrite, which share most of their observable properties but don't share the same hidden structure. The discovery of their different natures did not prompt any revision of the language. When we move from physics and chemistry to the biological sciences, more examples spring to mind. Whales were classified as mammals rather than fish, but things could have gone differently (LaPorte 2004).

Although most philosophers recognize an element of arbitrariness in the choices speakers make about the extension of natural-kind terms, they disagree on how to account for it. Some argue that natural kinds have an explanatory value and there is a variety of classificatory models which work for some purposes and not others: science does not discover how things are but provides different ways of representing them for different purposes (Dupré 1981). This is an anti-essentialist view of natural kinds.

Others claim that in some contexts scientists do carve things up in a way that is exclusively determined by the nature of the phenomena they study (e.g. names of elements in chemistry) but in other contexts there are factors external to science and these affect the way in which reference for a natural-kind term is fixed. For instance, rubies and sapphires have the same hidden structure but different names. Our vernacular use tracks the different surface properties of the compounds (red and blue), which we attribute to the different impurities present within their compounds, and not to their internal structure, which is the same. The surface properties have important implications for the market value of the stones. This is an essentialist view of natural kinds, but it recognizes that scientific interests play only a part in determining the use of natural-kind terms (Bird 2007).

Compare these views with Putnam's account: he argues that if tigers lost their stripes, they would not cease to be tigers, because having stripes is not what is essential to being a tiger. "X is striped" is just a description. Non-expert speakers may rely on it when they are asked to identify tigers but it does not figure among the essential properties of tigers. This account has important consequences for reference fixing: although most speakers recognize tigers by the way they look, and their having stripes of a certain kind and a certain color, the extension of the term "tiger" is determined by something else, whatever the essential properties of the individuals that were baptized as "tigers" are. If "tiger" is a natural-kind term, the essential properties of being a tiger will be identified by our latest theory of species classification in biology and other relevant empirical information about tigers.

 Exercise: If water lost all of its superficial properties but maintained its chemical composition, would it still make sense to call it "water"?

 Discuss: Does the answer to the previous question tell us something about how linguistic communities operate and evolve, or about meaning and reference?

There are other ways of accounting for reference fixing and naming. One option is to adopt a hybrid theory of reference: would it make sense to say that "tiger" refers to something completely different from what speakers now think it refers to, because a new scientific theory tells us so? Michael Devitt (1981) argues that there must be a descriptive element in the reference fixing of proper names, that is, there must be at least one thing that you know about the object you name for the causal relations between name and object to hold and play the all-important role of determining reference for generations to come. That is, a speaker needs to know, at the very least, what *kind* of thing is the object in question (in order to dub someone as "Homer" one needs to know at the very least that Homer is a man).

Can a similar approach work for natural-kind terms? If nothing of the original description associated with the substance dubbed "water" turns out to be accurate once chemistry has delivered the ultimate truth about the hidden structure of water, then it seems very counter-intuitive to claim that speakers have always been referring to water by their previous uses of the term. It is not sufficient to point at water and say "This is water" to baptize instances of water as "water," but some descriptive element should be added to the indexical (e.g. "This *liquid* is water"). If it is then discovered that water is not a liquid, one would have to conclude that in the past speakers had not been successful in referring to water by the term "water." Analogously, we could not fix the reference of "tiger" and ignore that tigers are animals (e.g. "This *animal* is a tiger"). If it was then discovered that the individuals called "tigers" were not animals, one would have to conclude that uses of "tiger" prior to this discovery and based on that original act of baptism did not refer.

Discuss: What determines the level of description of the qualifier (e.g. liquid, animal etc.)? What makes it the case that "tiger" would not refer if tigers turned out to be other than animals? Would it still refer if tigers turned out to be other than striped?

Some philosophers have accepted that in the scenario described by Putnam most of us have anti-descriptivist intuitions, but have raised concerns about the use of such a contrived example. After all, if two substances have a radically different hidden structure (e.g. H_2O and XYZ), then it is very implausible and unrealistic to suppose that there is no observable difference between them. At least in the chemistry lab, people on Twin-Earth behave towards what they call "water" differently from how people on Earth behave towards it, and form different beliefs about its properties and structure. But leaving these methodological concerns aside, what the thought experiment successfully does is pull out two distinguishable elements in any theory of meaning: what speakers think of when they use a term and what that term refers to in the world.

4.2. Implications of Descriptivism

Putnam takes the account of the meaning of natural-kind terms to be determined by a number of factors, among which the description commonly associated with the term ("Tigers are large feline animals with stripes") and the description of the extension of the term ("Tigers are animals with a certain DNA, etc."). When the theory about what tigers are changes, the stereotype does not change, and the extension of "tiger" does not change, because it has been fixed at the moment of baptism, but the theoretical description does.

For descriptivists, terms get their reference on the basis of the descriptions associated with them. Philosophers of science who defend a descriptivist account of the reference of theoretical terms often argue that communication among scientists before and after a significant theory shift becomes difficult, or even impossible, because the description of the theoretical terms has changed and with it their reference. For some theoretical terms whose descriptions are found to be completely inaccurate after the theory has been overthrown, we will have to say that they no longer refer. This claim has implications for both scientific realism and scientific progress. It is harder to defend the view that scientific theories are describing reality in a way that is at least approximately accurate and that each new theory is making progress with respect to previously accepted theories, if the theoretical statements of old and new theories cannot be compared.

We shall introduce the notion of meaning incommensurability here and then move on to address the question whether we are justified in believing that our current scientific theories are (at least approximately) true. We shall consider some of the implications of incommensurability for scientific progress in the next chapter.

4.2.1. Meaning incommensurability

Kuhn seems to imply that there is *no possibility* of understanding the language of one paradigm from the perspective of another, and, together with the possibility of understanding, opposed scientists also lack the possibility of communicating and comparing the formulations of and the solutions provided by their respective theories. He calls this the (meaning) incommensurability thesis "no common measure").

Kuhn seems to ground incommensurability on two theses: (1) the view that there is no pure observational language; (2) the view that theoretical terms change their meaning when theories change. He develops a criticism of the neo-positivist claim that all theoretical statements can be reduced to a shared observational language which is uncontaminated from theoretical assumptions. Through Carnap's idea that it is possible to translate any theoretical statement into a statement which contains only observational terms (see correspondence rules in the previous chapter), one could believe it possible to bridge the conceptual gap between two theoretical statements employing terms to which different descriptions are associated. The language of theory-free observation could be the intermediary and guarantee some minimal shared understanding. But the idea of a language that is purely observational and neutral with respect to theoretical interpretation is, for Kuhn, utopian.

Kuhn's analysis is inspired by examples in the history of science. Some terms have been redefined quite dramatically and to such an extent that the first uses of a term have almost nothing in common with the currently accepted uses of the same term. Consider the term "element" which has been introduced by the Ancient Greeks, and then used by the alchemists since the Middle Ages, and is now part of the vocabulary of contemporary chemists. For Aristotle, elements were the basic constituents of matter (earth, water, fire, and air) which could be turned into one another (e.g. water can become air) and were characterized by the properties of being wet or dry and hot or cold. The description of elements in chemistry today differs greatly from that offered by Aristotle, although they are still broadly conceived as the building blocks of matter. Elements today are more precisely characterized as those substances made up by just one type of atom, and scientists have discovered more than one hundred of

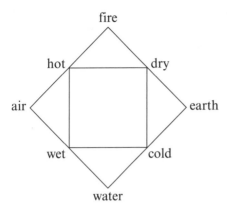

Figure 4.1
Elements and their
properties in
Aristotle.

them. Many of the properties of an element are seen as periodic functions of the number of protons within the atomic nucleus of the element, and the atomic number determines the arrangements of elements in a table which is meant to illustrate some of their periodic properties.

While we have continued to use the term "element," the description we associate with the term (what we mean by it), and the things we intend to refer to with the term (the referent) have changed dramatically across the centuries of scientific investigation, and similar examples abound in the history of science. But do these examples and the conclusions we can draw from them justify Kuhn's claim that theories are incommensurable?

If we were asked to translate the statement of a language into the statement of another radically different language, we would aim at preserving the most salient semantic properties and, given our aim, we probably would not achieve a satisfactory point-by-point translation. Translation can be generally problematic and locally impossible, as we experience every day when we speak a language we have learnt and which is not our native language. Still, to postulate that communication becomes impossible seems to many critics too extreme a conclusion. Partial understanding is often an aim that can be achieved with genuine effort. And in the case of theoretical statements belonging to a different theoretical framework, partial understanding is regarded as an essential requirement for rational theory choice. If the statements of two rival theories cannot be compared, then the choice between them cannot be made on the basis of considerations about their content (e.g. how much they can explain; how well they answer to potential objections; which implications they have).

Kuhn was right in emphasizing that theoretical terms may change their meaning when a theory is rejected and replaced. But the phenomenon is not confined to formal scientific disciplines. Even terms in everyday language are subject to often dramatic changes, in accordance with a change in background assumptions and beliefs. Let's think about another two examples, the everyday use of the term "witch" and the use of "combustion" in the chemical sciences.

Today the term "witch," if intended to refer to a person who has magical powers and practices sorcery, is generally thought to have no reference outside fiction. That is part of the reason why the term has a variety of metaphorical, as opposed to literal, uses. "She is a witch" can refer to a woman who is thought to be old and unattractive, or evil-looking. "She's bewitching" describes a person with charisma. But from the fifteenth to the eighteenth century the term "witch" indicated something else. Witches were people (usually women) who befriended the devil and were involved in a conspiracy to cause harm to other individuals or groups of individuals, and did cause harm by casting spells, bringing bad luck, and generally being responsible for plagues and catastrophes. When caught, they were often tortured until they confessed to having had contact with the devil and then killed. The term was thought to refer to actual people, and the description associated with it probably came from superstitious views about the possible causes of natural catastrophes and personal misfortunes, in conjunction with the observation of behavior that was regarded as unusual (e.g. possible cases of hysteria). What is interesting for our purposes here is that the change in background beliefs between those times and our times have determined a change in the description associated with the term "witch" and, even more significantly, a different answer to the question whether the term genuinely refers.

Exercise: Can you think of another case where the change in the description associated with a term together with the change in background beliefs have resulted in a term ceasing to refer to anything actual?

Now consider the term "combustion." Before and during the chemical revolution (which we shall describe in some detail in the next chapter), competing scientific explanations of combustion were available. Joseph Priestley thought of combustion as a process by which a burning object releases phlogiston, which is consequently absorbed by the surrounding air. For Antoine Lavoisier, combustion was a process by which a burning object combines with oxygen taken from the surrounding air. Following a descriptivist account of reference, whether the term "combustion" refers depends on the theoretical description associated with it. If combustion is what Priestley says it is, then it does not refer to any actual process (we now think) because there is no phlogiston. (A causal theorist would say that "combustion" always referred to something like the process which Lavoisier described, but had an inaccurate description associated with it before Lavoisier "discovered" oxygen.)

Kuhn is concerned with understanding. Can we understand the claims by sixteenth-century witch-hunters about the dangers of witches? Could Lavoisier and Priestley understand each other when they were talking about combustion? For the incommensurabilist the meaning of "combustion" is tied to the meaning postulates that the accepted chemical theory includes. The *holistic* conception of meaning the incommensurabilists embrace leads them to say that Priestley and Lavoisier thought they could understand each other, but actually they could not, because there was not enough common ground in their incompatible theoretical accounts of combustion, and the different meaning they attributed to "combustion" did affect the meaning of many, if not all, the other theoretical and even observational terms they were using. If we said that "the burning of an object" is a description that they could both recognize as a description of combustion, Kuhn would insist that they would have intended "burning" differently and that they could not have used and understood the term independently of the theoretical assumptions they were making about the nature of the process. "Combustion" for Priestley refers to something different to "combustion" for Lavoisier, as "Newtonian mass" has a distinct referent from "Einsteinian mass." Priestley was referring to something that (we now think) did not exist when he was talking about combustion, as he did when he was talking about phlogiston.

Discuss: Why should the association of a term with conflicting theoretical descriptions generate a failure of understanding? What are the assumptions Kuhn makes here?

4.2.2.   *Partial reference and imperfect translations*

The anti-incommensurabilist can reject descriptivism or meaning holism. One can insist that here is a meaning of combustion, as simple as "object burning," which is shared between rival theorists. Then there are incompatible views about how combustion takes place, but these do not necessarily prevent opposed scientists from partially understanding each other and having a grasp of what the rival theory wants to say. Some theoretical terms, included in a theory that has already proved to be false, are found by later scientists to fail to refer to something actual. The question whether, for instance, Newtonian mass exists is a controversial one. Mass as characterized by Einstein's relativity theory does not coincide with it and, since we have now rejected Newtonian physics and accepted relativity theory, we are led to think that there is nothing in the world *exactly* like Newtonian mass. On the other hand, the description of mass that Newton provided says something true about mass as we today intend it.

Hartry Field (1973) suggests that theoretical terms of already rejected theories only *partially* refer to objects in the world, and he uses just the example of "mass." The view is that it is to some extent indeterminate what "mass" as used by Newton did refer to. "Mass" in Newton's talk partially refers to *relativistic* mass and partially refers to *proper* mass.

When we use the word "mass" as Newton did we make a mistake not because the term we use no longer refers, but because it does not allow us to make all the relevant distinctions that we might need to make (Devitt 1997).

What about "phlogiston"? If we believe the oxygen theory, "phlogiston" lacks its referent entirely and there is no entity in the world that behaves in the way described by the phlogiston theory. We are forced to say that "phlogiston" has never had any referent at all. There is nothing we can pick out in the chemical world and identify as the (even partial) referent of that word. Scientists embedded in the theoretical framework shaped by a to-be-rejected theory may believe in the existence of objects that later on are shown to be merely theoretical "inventions".

Some have tried to refute Kuhn's incommensurability theory by using an argument that has as its target any instance of conceptual relativism (Davidson 1974). The relativist claims that there are different points of view but these points cannot be compared, because there is no common system of coordinates which allows us to locate the points. But if this is so, how can we recognize that the points are distinct in the first place? To deny the possibility of recognizing the difference by comparison is also to deny the difference itself.

This is a convincing rebuttal of those forms of relativism which suppose that there is a *given*, a content which has not been interpreted yet, and that different interpretations of it are incommensurable. If we know that the competing interpretations share a common basis (some neutral content to be interpreted), we have a ground for comparing them and for recognizing how they differ. But Kuhn's incommensurability thesis seems to be immune from this type of criticism. Kuhn does not believe in there being a common ground, a content which is not conceptualized, a language of pure observational character, and does not speak of different interpretations of shared data. He refuses the idea that there is an account of the data which is neutral with respect to various ways of interpreting it. The data themselves are different and the same observations can be accounted for in different ways before the stage of interpretation has been reached. The incommensurability invoked by Kuhn is just the absence of any common measure, of any system of coordinates in which the distinct points are located. We have to imagine a different system of coordinates for each point and no common plane in which they lie. The shift from a theoretical framework to another is a change of system as much as a change of point of view.

There is an alternative route to the dismissal of meaning incommensurability. First, we can concede that a point-by-point translation is not only often impossible, but that, even when it seems possible, it is a serious challenge and requires the translator to make difficult and controversial choices. Some loss of meaning is most likely. Partial incommensurability, then, can be experienced and it is part of the shift from one theoretical language to another: there may be (local) failures of understanding due to the use of the same term to refer to entities which are ascribed different properties. And yet, the admission that accurate translation between theoretical languages is fallible does not compel us to embrace the incommensurability thesis in its full strength. Rival theorists understand (at least to some extent) which differences characterize their positions (or they could not engage in any meaningful debate) and this observation alone makes incommensurability thesis implausible, insofar as the incommensurabilist insists on the very possibility of communication being compromised and on the inevitability of misunderstandings. The thought that in general, as opposed to in particular historical circumstances, scientists are systematically deluded when they believe they are successful in communicating with rival theorists, is also an implausible thesis.

Let's consider a couple of examples where understanding seems within reach. Recall our previous discussion of the term "element" and how the theoretical description associated with it changed from Aristotle to contemporary chemistry via centuries of alchemical tradition. When we are thinking about statements containing the term "element," some of the claims that a contemporary of Aristotle would have judged as true (e.g. "Water is an element") are now false (e.g. water is not one of the *basic* building blocks of matter). But we understand what Aristotle meant by calling water an element, although this understanding comes only at the price of learning about the background beliefs held at his time about the constitution of matter and motion and about the implications of something being classed as an element for physical and metaphysical theories.

Here is another example. Henri Poincaré (1902, 2003, p. 42) attempted to create a peculiar dictionary to translate crucial terms of classical Euclidean geometry into the terms of a non-Euclidean geometry, Lobatchewsky's hyperbolic geometry. The development of new geometries in the first half of the nineteenth century meant that Euclidean geometry was no longer the only way to represent space. Moreover, the claim that Euclidean geometry was true of our empirical world was also put under pressure, given the conception of space introduced in Einstein's relativity theory. Einstein's conception of space is not entirely compatible with Euclidean geometry, which turns out to be only an approximation to the geometry of the actual physical space. In Lobatchewsky's geometry the fifth postulate of Euclidean geometry is rejected, and this is the postulate according to which two converging straight lines must eventually intersect in the direction in which they converge.

How would the terms used in Euclidean geometry be translated into terms of Lobatchewsky's hyperbolic geometry? Poincaré claims that, if a dictionary of geometrical terms existed, it would be possible to translate not just terms, one by one, but even theorems. What we call "space" in Euclidean geometry is the portion of space situated above the fundamental plane in hyperbolic geometry, and Lobatchewsky's theorem according to which the sum of the angles of a triangle is less than two right angles is translated as follows: "if a curvilinear triangle had for its sides arcs of circles, which, if produced, would cut orthogonally the fundamental plane, the sum of the angles of this curvilinear triangle would be less than two right angles."

This is an interesting case in which an imperfect translation between theoretical formulations produced within rival paradigms can be attempted. It is not an elegant point-by-point translation and all the conceptual complexities of the comparison are reflected in the linguistic effort of transmitting ideas with inadequate tools. But it shows that, sometimes, we can save inter-theoretical communication and commensurability by recognizing the challenges that conceptual change creates and build tentative bridges with the resources that are available.

 Exercise: Can you think of another example of a partially successful translation between statements belonging to competing theories?

4.3.   Realism

Philosophical realism is a view about what there is and what we can know, and thus it has both an ontological and an epistemological component. According to the ontological thesis of a typical realist, there are mind-independent objects and properties in the world, objects and properties that would be there even if there were no mind to think about them. According to the epistemological thesis, we can have access to these objects and properties, and our representations of them, though fallible, are not systematically misleading. Often the realist position is characterized in terms of our perceptual experience being a reliable guide to what is independent of our experience. The realist would answer affirmatively to questions such as: Does the car that I see in front of me really exist? Does it appear to me (roughly) as it really is?

In everyday life we rarely stop and consider whether our experiences are veridical. But there are circumstances in which we are in doubt. Faced with the possibility of an illusionist's trick, I might consider the possibility that the car I see in front of me is just a sophisticated hologram or a cleverly obtained mirror image. Maybe there is a car somewhere, but it is not in front of me. Skeptics invite us to generalize from these rare occasions and ask us to imagine that similar doubts might affect the bulk of our experiences. But realists reply that our experience of objects and properties in the world gives us reasons to believe that those objects and those properties exist, independent of our perception of them, and our largely successful causal interactions with such objects and properties give us reasons to believe that the way in which we experience them is not systematically misleading.

Although realism is the default, commonsense position, we can easily imagine skeptical challenges that would insinuate doubts about

what there is and what we know. A typical scenario in which the information we get through perception is not a guide to what the real world
is like is that of brains in a vat. Hilary Putnam (1981, 1999) suggests that
the perceptual experiences we have might come not from us interacting causally with objects and properties in the world, but from our
brains having being removed from our bodies, kept alive in the lab of a
mad scientist and receiving electrical impulses that provide stimulation indistinguishable from the stimulation our brains would receive
when perceiving objects and properties in the world. What makes this
hypothesis so interesting (and similar hypotheses, such as Descartes'
hyperbolic doubt or the possibility of a comprehensive computer simulation as in the *Matrix* movies), is that if we were just brains in a vat, we
would have no way of knowing it. Based on our experiences alone, we
are not able to discriminate between the realist hypothesis that our
experiences are veridical and caused by the interaction of our bodies
with other objects and properties in the world and the skeptical
hypothesis that our experiences are just the result of a mad scientist
stimulating our disembodied brains (for a recent discussion of this
point, see Pritchard 2005 and Goldman 2007).

4.3.1. Realism in the philosophy of science

The discussion of *scientific* realism concerns the capacity that our
current scientific theories have to describe and explain reality and the
structure of the debate between realists and anti-realists is not dissimilar from the structure of the debate about skepticism that we summa-
rized above. The force of the anti-realist hypothesis is in the thought
that, even if our current scientific theories did not correctly map onto
real objects and properties in the world, they would probably still work
well enough, and we would have no way of discriminating between their
being (approximately) true and their being just empirically adequate.

In the philosophy of science, questions about realism are most
frequently phrased in terms of the ontological status of both the unobservable entities and relations between events that are posited by our
current scientific theories. Do electrons really exist? Is inflation real?
Here is the challenge to the realist position: nothing would justify the
step from *accepting* a scientific theory that can efficiently predict the
phenomena within a given domain to *believing* that the entities and
relations posited by that theory really exist. Notice the analogy with the
previously discussed skeptical scenario: no amount of perceptual
information we get about the world can tell us that we are not brains in
vats or trapped in the matrix. No amount of confirmatory evidence we
gather for any specific theory can tell us whether the theory is more
than just a useful tool for the prediction of the events that we want to be
able to control.

In the rest of this section, we shall review some of the most common
challenges to scientific realism and sketch some alternative positions

to the realist one. The target-theses are the claims that current scientific theories are true, and that the theoretical entities and relations they posit truly exist.

4.3.2.   Arguments against realism

The pessimistic meta-induction
If all previously accepted scientific theories have been found to be false, why should we believe in the truth of our current theories? This argument was made popular by Hilary Putnam (1978). He starts from the assumption that most current scientific theories are true. He then argues that most past scientific theories must be false, because they differ from and have been replaced by the current ones. By induction on past theories, we conclude that most current scientific theories will turn out to be false too. This conclusion conflicts with our initial assumption. The point of the argument is to make us realize that we are not justified in holding that our current scientific theories are true.

 Exercise: Why is this argument a "meta-induction"?

Larry Laudan (1981) puts forward an argument with a similar structure, as it consists in assuming first that the empirical success of a theory is an indication of its truth and then finding that this assumption leads us to a contradiction. If most of our current theories are true, then most past scientific theories are false, because they have been replaced by and differ from the theories we currently accept. But many of these past theories *were* empirically successful. Thus, the fact that a theory is empirically successful cannot, after all, be regarded as an indication that the theory is true. We can apply this consideration not just to past theories, but to current theories too. Their empirical success is not a good enough reason to believe that they are true.

Laudan's version of the argument has been much discussed in the literature on realism and has received two types of criticism. Some authors believe that it is not valid, and therefore does not present a real challenge to realism. Other authors have taken the argument seriously and have made it their project to reject one of the premises in order to defend a version of realism.

In the rest of the chapter we shall review some attempts to develop moderate forms of realism that do not seem to be as vulnerable to Laudan's challenge. But first we should address the concerns about the validity of the argument. It has been pointed out that some versions of the pessimistic induction can be charged with the "turnover fallacy" (Lewis 2001; Lange 2002). The following example might help understand the main gist of the objection: the fact that the membership of a board of directors has been subject to many changes in the past does not imply that all the current members of the board will be soon replaced. It may be that some members have occupied their posts for many years, and are not going to be replaced in the near future. From the fact that many changes occurred in the membership of the board one can infer that more changes will occur in the future, but one cannot predict when each member is going to be replaced.

On the basis of evidence from the history of science, we might come to believe that the theories we currently accept will be regarded as false in the future and replaced by other theories, in spite of the fact that they are empirically successful right now. But this judgment is not justified for all current theories, as it may be the case that some of them have been accepted for quite a long time and will continue to be accepted in the future. The argument would need as one of its premises the claim that, *at any one time in the past*, most of the accepted theories were false (by our current standards). But this is a much harder claim to justify on the basis of historical evidence.

In response to this objection, one could argue that, in order for the argument to be a challenge to realism, the conclusion does not need to be that most of our current theories will be replaced some time soon (Saatsi 2005). Rather, all the challenge requires is to accept the claim that empirical success is not necessarily a guide to truth, and this claim is compatible with the possibility that all of our current scientific theories are both empirically successful and true, and are never going to be replaced in the future. However, the argument would be somehow less compelling if the inductive evidence did not lead us to believe that what happened to successful past theories is likely to happen to our current theories.

 Discuss: Which is the weakest premise in the argument for the pessimistic meta-induction?

Underdetermination

Suppose that there are two theories that are empirically equivalent, that is, that do not differ in the conclusions they reach about *observed* objects and relations. In terms of prediction, the two theories deliver the same results, but those results are explained differently by the two theories, because different *unobservable* entities and relations are posited.

The debate between supporters of the Copernican and the Ptolemaic astronomical systems in the sixteenth century exemplified this situation – before new data became available in favor of the heliocentric system. The theories were both compatible with the observations of the motion of planets and moons, but the way in which the trajectories of these bodies were calculated were very different.

Another example could be the explanation of psychopathologies: before neuroscience could provide evidence of brain damage for at least some of the conditions, there was a lively debate between psychodynamic accounts and cognitive neuropsychological accounts. Consider the Capgras syndrome as an example. This is the delusion people suffer from when they claim that a loved one has been replaced by an impostor. According to psychodynamic accounts, the conviction that the person thought to be (almost) identical to the spouse was not the spouse is due to an attempt to reconcile negative feelings towards the spouse with the sense that it would have been wrong not to love her. According to the cognitive neuropsychological account, the condition is due to brain damage to the system of recognition of familiar faces. Today the latter model of explanation is the dominant one, but, before evidence of brain damage in subjects with Capgras was available, the evidence alone did not discriminate between the two hypotheses.

The observation that a theory has empirically equivalent rival theories is called "inductive underdetermination" (Okasha 2002) or "weak underdetermination" (Devitt 2005). A formulation of the inductive indetermination thesis is that no *actual* evidence at the present point in time can help us discriminate between two or more theories that are empirically adequate. This does not rule out that in the future new evidence could tip the balance and help us make a choice between those theories (as it did happen during the Scientific Revolution when the Ptolemaic system was abandoned in favor of the Copernican one and when data in neuroscience supported the cognitive neuropsychological model of delusions).

A more radical thought is that no *possible* evidence can ever help us decide between two or more theories, because no matter how much more evidence we can gather, the theories will be indistinguishable on merely empirical grounds. This latter type of underdetermination, often called "strong underdetermination," has been constructed as a challenge to some versions of realism (Duhem 1969; Van Fraassen 1980; Putnam 1983). But what is the challenge exactly? The thought is that, if there are two empirically equivalent theories, the realist is not justified in asking us to commit to the truth of one of the theories or to the existence of the entities and relations posited by one of the theories. The appeal to underdetermination forces us to review the relation between theory and evidence, because the underdetermination thesis suggests that by evidence alone we cannot tell whether a theory is true. All we can establish by considering the evidence is whether the theory is empirically adequate.

Here is a (very simplified) example taken from studies in cognitive ethology and comparative psychology. Since the '70s primatologists have been interested in whether primates are able to interpret the behavior of others (e.g. their conspecifics or a human trainer) on the basis of the ascription of unobservable mental states. This capacity is often referred to as a "theory of mind." It is no surprise that chimpanzees can interact in a group and that this requires some coordination. But what is not clear is whether they can think in terms of other individuals having beliefs, desires, emotions, and so on.

Many studies have been devised to determine whether the behavior of primates is (a) behavior that responds only to the observation of the behavior of others or (b) behavior that responds to the observation of the behavior *and* mental states of others.

There are two dominant hypotheses in this debate: (1) primates respond to the behavior and mental states of others in some circumstances (Tomasello and colleagues); and (2) primates do not respond to underlying mental states of others (Povinelli and colleagues). The question that people in the debate have been discussing recently is whether the evidence available so far can discriminate between these two hypotheses. Can we legitimately conclude that one hypothesis is better supported by the evidence?

If we were to consider the large body of ever-growing evidence on the behavior of primates, it is unlikely that the two hypotheses would consistently receive equal support, but we shall focus on one very crucial experiment. With respect to that experiment we have an instance of underdetermination.

The experiment is part of a series of studies on attention and gaze-following. It attempts to establish whether chimpanzees can distinguish between trainers who can see them and trainers who cannot see them. When chimpanzees want something, they use a natural "begging gesture" which is a visual signal. If chimpanzees were placed in an enclosure with two people, one facing them and one turned the other way and unable to see them, and were systematically directing their begging gesture to the person who was facing them, this would be regarded as evidence that they understand "seeing." The experiment offers this result, that chimpanzees approach the trainer who can see them when they ask for food. So we have prima facie support for hypothesis (1) by Tomasello and colleagues, that primates can respond to the mental states of others (the state of seeing, at least) in some circumstances.

But supporters of hypothesis (2) claim that the experimental result could be explained equally well by the fact that chimpanzees know from past experience that organisms facing them have a greater probability of reacting to their begging gesture than organisms that are not facing them. So there can be a good explanation of the experimental results that does not need to assume anything at all about the capacity chimpanzees might have to respond to the mental states of others. They could just be responding to others facing-forward.

This is a case in which the experimental result taken in isolation from other experimental results does not seem sufficient to discriminate between the two hypotheses and can be explained equally well in terms of primates responding to behavior and mental states or primates responding to behavior alone. This does not mean that one hypothesis cannot be preferred to the other, but that the criteria for this preference will not be evidential criteria. If we believe that parsimony is a virtue in theoretical explanations, we might accept a general recommendation for theory choice, according to which, when evidence does not favor one hypothesis over the other, we should prefer the hypothesis that provides us with the simpler explanation or commits us to the existence of fewer objects. In this case, it is arguable that the hypothesis defended by Povinelli and colleagues will be selected as the most parsimonious, as it does not need to commit us to the view that primates can have a theory of mind. But even this assessment is open to debate.

 Exercise: Can you think of another example to illustrate the weak underdetermination thesis?

Duhem–Quine thesis

There is another thesis which suggests that by evidence alone we cannot even tell whether a theory is false, contrary to the aspirations of Popper's falsificationism. This is the so-called Duhem–Quine thesis. The thesis is based on the observation that, when we are testing a theory, we are not testing it in complete isolation from other hypotheses. If the experiment delivers a result that conflicts with the prediction made on the basis of the theory and the auxiliary hypotheses used to devise the experiment, then the scientist is left with a difficult choice: she can either reject the theory or challenge the auxiliary hypotheses. What is rejected as a result of the experiment is not *determined* by the empirical evidence, but it remains at the discretion of the scientist.

Think about the attempt to falsify a theory T_1. It is impossible to design an experiment whose result can be predicted on the basis of T_1 alone, as, for instance, the instrumentation used will also rely on other theoretical assumptions (h_1 and h_2). What the scientist is putting to the test is not only one theory, but the conjunction of the theory with those hypotheses that are involved in the design of the experiment. If the observation does not correspond to what the scientist had predicted, the conjunction of the theory and the other assumptions must be false, but the scientist does not know which of the conjuncts is responsible for the empirical failure:

Premise 1: $T_1 + h_1 + h_2 \rightarrow O$

The conjunction of the theory and the assumption allows us to predict that a certain event will occur.

Premise 2: Non-O

The predicted event does not occur.

Conclusion: Non- $[T_1 + h_1 + h_2]$

Then we have to conclude that the conjunction of the theory and the assumptions is false. But is it the theory that needs to be rejected or one of the assumptions? How can we tell?

Here is an example of how this strategy can be applied. In cognitive psychology one project is to find out whether humans have reasoning competence which matches norms of correct reasoning. This research project involves testing human reasoning capacities by asking research participants to solve tasks whose correct solution is determined on the basis of the norms of logic, statistics, probability, or decision theory. If most research participants fail the given reasoning tasks, then the conclusion is that their reasoning competence does not match the norms of correct reasoning.

In the experiments, this is the thesis cognitive psychologists want to test: (T1) *human reasoning competence conforms to the norms of correct reasoning*. These are two of the assumptions that need to be made in order to derive failure of competence from failure to solve the tasks: (h1) *the norms of correct reasoning are given by logic, statistics, probability, and decision-theory;* (h2) *errors made in the attempt to solve the given tasks are not performance errors.*

Suppose the experiment takes place and most research participants fail to solve the tasks ($T_1 + h_1 + h_2 \rightarrow O$ and non-O). Does this observation mean that they do not have a reasoning competence which matches the normative standards of correct reasoning? No, it just means that the conjunction of the thesis and the two assumptions is false (non- $[T_1 + h_1 + h_2]$). Which conjunct we reject is an open question (and this open question gave rise to the Rationality Debate in cognitive science). We can reject the view that reasoning competence is inadequate by rejecting the view that logic, statistics, probability theory and decision-theory provide norms of good reasoning, or by arguing that the mistakes were due to performance errors.

 Exercise: Can you think of another example to which the Duhem–Quine thesis can be applied?

Duhem is not entirely pessimistic about the consequences of his thesis for the reliability of the scientific method and argues that scientists can use their "good sense" to make the necessary decisions. Quine (1951) develops a much wider-ranging holistic thesis according to which it is our knowledge in its entirety that we put to the test whenever we interrogate nature. Our belief in one single hypothesis can never be shown to be false by experience on its own, because it is the whole web of beliefs that faces confirmation or disconfirmation every time we experience a new phenomenon. Then we have the choice of rejecting one or more beliefs in the system and, according to Quine, we should be conservative and give up the belief whose rejection brings fewer changes to the rest of the beliefs we have. For Duhem, to say that the Ptolemaic system is false does not mean that it has been falsified by experience, but that it does not fit the other accepted hypotheses as well as the Copernican system. For instance, it is not compatible with Newtonian physics.

To sum up, a challenge to realism could be made on the basis of the view that empirical evidence alone cannot tell us whether a theory is true (underdetermination thesis) or whether it is false (Duhem–Quine thesis). What we can say about the hypotheses we currently accept is that, in conjunction with other hypotheses, they allow us to make reasonably accurate predictions. Realists typically respond in two ways to the challenges raised by the underdetermination thesis: either they deny that there can be incompatible theories that are nonetheless empirically equivalent; or they concede that there can be empirically equivalent theories that are incompatible, but argue that a choice can be made between them on grounds other than empirical adequacy.

 Discuss: Can there be a form of realism which is immune from the underdetermination thesis?

4.4.   The Realism Debate

Before assessing the arguments in favor of realism, let's see what the alternative positions are. We shall introduce three options: *instrumentalism*, characterized by the semantic view that theoretical statements are not truth-assessable because the theoretical terms they contain do not refer; *constructive empiricism*, whose main claim, of an epistemic nature, is that we cannot be justified in regarding our scientific theories as true rather than empirically adequate, given the evidence available to us; and *the natural ontological attitude*, which is based on the rejection of both realism and anti-realism and advocates the view that, if scientific theories are true, they are true in the same way in which ordinary sense perception is.

4.4.1.   Alternatives to realism

Instrumentalism
The view that characterizes instrumentalism about science is that theories are instruments that we deploy in order to predict events. Against the realist, the instrumentalist claims that, whereas observation statements are true or false, our theories are neither true nor false. We accept theories not because they are true, but because their predictions are accurate. And we can legitimately assess the accuracy of their predictions because we can see whether the observation statements that derive from our theories are true.

This position relies on a distinction between theoretical statements (that are neither true nor false) and observation statements (that are true or false). What explains this difference? Why cannot the truth of theoretical statements be assessed?

According to the instrumentalist, all the statements that contain theoretical terms are neither true nor false because theoretical terms *do not refer*. Instrumentalists argue that when the meaning of a term is not associated with anything observable, but its meaning depends exclusively on a theoretical description, then the term does not have a reference. An example of an extreme instrumentalist position (*eliminativist view*) is that of Ernst Mach who believes that physical objects are nothing but bundles of sensations. On this view, only sensations are real. Science has the purpose to postulate convenient fictions which allow us to track the way in which sensations relate to one another, but claims about atoms and other unobservable entities have no truth-assessable content.

 Exercise: Does the plausibility of instrumentalism depend on a specific account of reference?

There are several brands of instrumentalism, and according to Pierre Duhem's version of the view, scientific theories do not aim to provide any explanation of reality (*anti-explanationist view*). Rather, its aim is to "save the phenomena," be compatible with the available data. Explanation is reserved to metaphysics. According to him, scientific hypotheses are not statements about the nature of reality and do not have a truth-value. Thus, they should not be assessed on the basis of whether they correctly capture how reality is, or whether they are true, but only on the basis of whether they are *convenient*. Convenience is determined by the consequences of these scientific hypotheses fitting the data. Recall the examples of weak underdetermination we explore in the previous section.

We saw that there is a debate between accounts of the behavior of primates. These accounts differ because they cannot agree on whether primates have the capacity to think about the mental states of others. According to the instrumentalism Duhem defends, it does not matter which account we adopt as long as we can successfully predict the behavior of primates on the basis of that account. This is because science is not supposed to provide an explanation of why the chimpanzee uses a certain begging gesture with greater frequency when the trainer is facing forward.

Constructive empiricism
Constructive empiricists agree with instrumentalists about the purpose of science: scientific theories are not supposed to describe how things really are, but to assist experimentation by allowing scientists to formulate clear questions and inform the design of experiments that can answer those questions. If this is what theories are for, all we can say about them when they are not refuted is that they are empirically adequate, but not necessarily true.

The difference between instrumentalism and constructive empiricism is that according to the former theoretical statements are not truth-assessable and according to the latter, theoretical statements are truth-assessable but we lack sufficient justification for claiming that they are true.

Why don't we have enough evidence to support the truth of the theories we accept? The reasons to believe in the non-observable entities that a theory postulates lie in the experimental findings that can be accommodated by using that theory. Saying that the theory is true, where truth means something over and beyond empirical adequacy, would be to claim that, outside the practice of science where that theory is used, we are committed to the existence of the entities postulated by the theory. But this does not even make sense, because there is no way of even describing those entities if not by using the vocabulary of the theory that postulates their existence (Van Fraassen 1980).

The acceptance of a theory in this picture does not depend on truth, or approximation to the truth, but has a pragmatic dimension. Once competing theories have been found to be both consistent and empirically adequate, other criteria will recommend which one should be accepted, on the basis of the pragmatic role they can play in the practice of science.

Think about our previous example of the explanation of psychopathologies in terms of psychodynamic theories or within the cognitive neuropsychological framework. Both types of explanation require positing non-observable entities, such as repressed desires and unconscious motives, or damage in the frontal lobe of a human brain, which can be "observed" only via complex and theoretically laden instrumentation (e.g. MRI scans). According to constructive empiricism, we would not be justified in claiming that either of these explanations is true, or is the correct one, on the basis of the empirical evidence alone, as the only claim that can be justified on empirical grounds is a claim of empirical adequacy. We shall choose the account that fits better the observations we can make about delusional subjects' behavior, and if they are both acceptable on those grounds, we shall make a choice on the basis of other criteria (e.g. whether the explanation fits with a materialistic conception of the mind that we have independent reasons to prefer).

Exercise: Would the very idea of constructive empiricism be compromised if the distinction between observable and non-observable entities were undermined?

NOA: Realism is just metaphysical foot-stamping

Arthur Fine claims that arguments for realism and anti-realism should be resisted and defends an intermediate position that he calls the *Natural Ontological Attitude*, whose main thrust is that we should regard scientific truths in the same way as we regard everyday truths coming from sense experience, because there is nothing about scientific theories that makes them better at telling us what the *deep nature* of reality is.

He considers two arguments for realism and argues that both are weak. One is that it cannot be a coincidence that at any one time in the history of science only a small number of theories are taken to be plausible and that the successors of the theories we had in the past are similar to them (the *small handful* argument). For the realist this is evidence that past and present theories approximate the truth and are not just wild guesses made with the purpose to accommodate the evidence in any coherent way. But Fine replies that anti-realists can also explain the coincidence: the elements of the previously accepted theory that are responsible for the delivery of accurate predictions are preserved and new hypotheses are added to the empirically successful portion of the theory in order to make novel predictions. This method can increase empirical adequacy without necessarily leading to the truth.

The other realist argument Fine dismisses is that sometimes successful predictions are made on the basis of the conjunction of two accepted theories. The realist argument suggests that this phenomenon can be explained easily if we believe that both theories are true, as their conjunction will be true as well, but cannot be explained satisfactorily if we take the two theories to be only empirically adequate, as there is no reason to believe that their conjunction will also be empirically adequate. Fine is not convinced by this argument. Although the conjunction of true hypotheses will also be true, it is overwhelmingly plausible (given considerations about the pessimistic meta-induction) that our theories are just *approximately true* rather than true. If this is the case, then we have no reason to believe that the conjunction of approximately true hypotheses will also be approximately true. Fine's conclusion is that a sensible realist cannot successfully deploy this argument.

 Discuss: Does Fine's position differ from constructive empiricism?

4.4.2. Defending realism

Realism is often presented as the commonsense, default, uncritical attitude to adopt. But arguments for full-blown realism have been found unsatisfactory because they do not seem to support the claim that current theories are true or approximately true rather than just empirically adequate. Under the pressure of anti-realist challenges, realists have developed sophisticated defenses of their position.

Consider the classic *No-Miracles* argument. The simple structure of this argument is an inference to the best explanation. We want to explain the overwhelming and uncontroversial success of science. Realism provides an explanation for the success of science: our current scientific theories are true and their theoretical terms refer, and this is why the predictions that we make on the basis of those theories are accurate. But it is not entirely clear why the claim that scientific theories are true should provide an explanation for the success of science that is not already available by merely accepting those theories (Bird 1998). To accept a theory does not necessarily involve the belief that it is true, but the belief that it receives support from the available evidence and can be fruitfully used within a given research program.

Can there be a form of realism which concedes something to the anti-realist arguments and yet resists their conclusions? We are going to review two possible approaches to moderate realism: structural realism and internal realism.

Structural realism

The arguments against realism target the claim that scientific theories can describe the *nature* of reality. Such arguments gain credibility from the observation of the failure of past scientific theories. John Worrall (1989) has argued that there is a version of realism that is open to those who struggle to believe in the existence of electrons and quanta, and that can survive the challenges of the anti-realist. This is structural realism, the view according to which theories in mature science capture the *relations* among phenomena they attempt to explain, even when they fail to capture the *nature* of those phenomena. According to Worrall, this view is what many philosophers regarded as anti-realists or instrumentalists (e.g. Duhem and Poincaré) really had in mind all along.

The view relies on the distinction between structure and content: content-wise, it is true that past theories have turned out to be false, as some of the theoretical entities they posited are now believed not to exist; but what has been retained of past theories is the formal relations (often expressed in mathematical terms) among the entities that were posited (Psillos 1999). This is compatible with a weak version of the No-Miracles argument, as it would partially explain why even past theories that have been rejected were to some extent empirically adequate, but would prove less than the No-Miracles argument, as it would not be committed to the claim that the theoretical terms of current scientific theories successfully refer.

The example used by Worrall is that of the theory of light and the shift from Fresnel's *particle* theory to Maxwell's *wave* theory. Fresnel thought that light was conducted via an elastic solid, whereas Maxwell believed that it was carried by waves within an electromagnetic field. Although they disagreed on the nature of light, they formulated laws that are formally very similar, which might explain why Fresnel's theory could predict accurately many observable optical phenomena. The thought is that by being structural realists we can explain the continuity that we detect even among theories that precede and follow a scientific revolution, and therefore account for the fact that the success of predictions made on the basis of false theories is not a miracle.

The problem with this account is that we need a good justification for the claim that in a scientific theory it is always possible to distinguish its content (which may not be preserved in future theories) and its structure (which is more likely to be preserved).

 Exercise: Are there other examples of theory-shifts where the change can be seen as a change in content rather than a change in structure?

Internal realism

Hilary Putnam (1983) provides another compromise-solution to the radical opposition between realists and anti-realists by arguing that talking about objects existing independently of our conceptual ways of carving up nature is empty. Putnam wants to distance himself from full-blown (metaphysical) realism, because he believes that it would lead to attempting to reduce the existence of middle-sized objects, such as trees and chairs, to more fundamental objects described by, say, contemporary physics and, as a consequence, to claiming that middle-sized objects are not real. On the other hand, he rejects instrumentalism and constructive empiricism because he believes that chairs and electrons have the same status, and that the entities posited by accepted theories are no less real than observable entities.

The questions: "How many objects exist?" and "Which objects exist?" cannot be answered independently of our concepts of "object" and "existence," and in this sense there are different versions, different descriptions of reality. Our answer to how many objects there are will vary according to whether we are counting molecules or pieces of furniture: given that *we* determine what an object is, it makes little sense to ask whether objects exist independently of *us*. For the scientific realist, only fundamental objects as described by physics exist, and the objects visible to us are just projections of our minds; for the anti-realist, only what we see is real and the entities posited by physical theories to explain observable events are fictions.

For Putnam both what we see and what we cannot see is real, within the conceptual scheme in which we operate. But he also wants to distance himself from relativism. To say that our concepts determine the answer to the question how many objects there are does not mean that we agree by *convention* on one answer and that our answer is as good as any other. Rather, our concepts determine our way of counting objects but the answer to the question how many objects there are is still to be "discovered."

 Exercise: Identify similarities and differences between Fine's NOA and Putnam's internal realism.

Summary

In this chapter we have considered some issues that emerge from the language of science, and in particular how natural-kind terms get their meaning and reference, and whether theoretical terms refer. These issues transcend semantics: the causal theory of reference invites us to see science as discovering the essence of the fundamental blocks of reality, whereas the descriptivist theory is more interested in the way in which our concepts are a guide to reference. Descriptions associated with natural kinds need not disclose any essential properties. Similarly, the question whether theoretical terms get to refer has implications for the issue whether scientists endorsing rival theories can successfully communicate and for the claim that scientific theories grasp the real, deep nature of reality.

If theoretical terms do refer, then theoretical statements containing them are either true or false. If they are true, they describe how things are and are more than a useful predictive tool. But the anti-realist challenges the metaphysical assumption that science is the route to our knowledge of reality and argues that theories play a more modest role. We saw that in between full-blown realism and anti-realism there are a range of moderate positions that tend to relativize claims about reality to facts about our conceptual scheme, or suggest that scientific theories capture how things are, but no better than our ordinary talk of chairs and table does.

The relationship between the language and reality is a very complex one and we have only scratched the surface. But this brief introduction to the existing literature on scientific realism is necessary to understand the issues raised by the analysis of scientific change and its rationality, and to assess how they contribute to the definition of the role of science in society.

Preview of future attractions

The way in which we view the language of science, especially the issue of the reference of theoretical terms, and the debate on the status of scientific theories are crucial to accounts of scientific change and progress. In the next chapter, we shall address the question whether theory choice is based on objective criteria and whether progress can be regarded as cumulative in the face of radical conceptual changes in the history of science.

Issues to think about

1. Can the theoretical statements of Newtonian physics be reformulated within Relativity Theory?
2. For which types of words is descriptivism better suited?
3. Is there a substantial difference between the positions of a constructive empiricist and a moderate realist?
4. Could we be essentialist about social kinds? What fixes the extension of social-kind terms like "race," "famine," "feudalism," "gender," "revolution"?
5. Do you think that thought experiments (such as Putnam's Twin-Earth) have a useful methodological role to play in providing justification for philosophical positions? Identify some advantages and disadvantages of the use of thought experiments.
6. How is internal realism different from conceptual relativism?

Further resources

The literature on the meaning and reference on natural-kind terms is vast and will lead you to explore both the epistemological and metaphysical dimensions of the debate between descriptivism and the causal theory of reference. A good place to start is the seminal paper by Hilary Putnam "The Meaning of 'Meaning'" where the Twin-Earth scenario is discussed in some detail. In Kripke (1980) you find the philosophical framework behind the idea of rigid designation and baptism which will be useful for an appreciation of essentialism. How these ideas are applied in science can be seen in the work of Bird (2007), LaPorte (2004), and Dupré (1981).

For more details about the many ways in which the idea of incommensurability has been developed, you can start by reading Kuhn (1962, 1970, chapter 11) and Feyerabend (1975). Critiques of this idea as applied to the change of reference of theoretical terms can be found in Devitt (1979) and Field (1973). Implications of incommensurability for progress will be discussed in the next chapter.

For an excellent introduction to the debate between realists and anti-realists in science, see Bird (1998, chapter 4) and Papineau (1996). Van Fraassen (1980) is the *manifesto* of constructive empiricism and Fine (1984) provides an interesting compromise between realism and anti-realism. For a realist perspective, see Boyd (1990).

If you want to learn more about the so-called Duhem–Quine thesis and its implication for the refutability of theories and scientific realism, you can start from a collection of papers on the topic, Harding (1976), and read Lakatos (1978, appendix to section 1).

5 Rationality

The success of science is often celebrated as the most striking of human achievements. However, we saw that it is difficult to point at what is special about science: inductive generalizations which underlie the practice of science are fallible; the method used by the sciences cannot be easily accounted for in a unique and distinctive way; a scientific theory does not need to be true or to present us with an accurate description of reality in order to receive empirical confirmation, to be deployed in explanation, or to function as an efficient predictive tool. In the light of these debates, the question whether we are justified in ascribing to science a privileged status among other human practices becomes even more pressing. But before we can provide an answer to this question, we have two more issues to tackle: Does science evolve in a progressive way? Is change in science guided by rational principles?

Even if we are realists about scientific theories and believe that the entities and relations posited by current scientific theories really exist, the nature of progress in science forces us to recognize that theories previously accepted are no longer regarded as true, and that they posited the existence of entities or relations that we now think never existed. How can we be confident that our current theories are really better than our previous ones, rather than merely different? Can the choice of which theory to adopt be supported by rational argument? When a scientific community overthrows a theory and embraces another, this happens often through a process of gradual change that philosophers of science have long attempted to understand by using competing models.

The issues of whether scientific change is rational and of whether progress is cumulative have been the object of the traditional debate between rationalists and historicists, and have been addressed together with other issues, such as the reference of theoretical terms we no longer use, the purpose of science, the capacity of scientists to communicate effectively with those who defend competing theories, and the plurality of styles of reasoning that characterize human inquiries into nature. If we think that the theory we now accept is closer to the truth (or more empirically adequate) than the previous one, and that it can explain more facts in a more satisfactory and comprehensive way, we will tend to see the shift from the previous theory to the current one as an instance of progress, based on rational principles of theory choice.

Rationalists depict change exactly in this way: the scientific community goes forward on the basis of sound argumentation supported by solid empirical evidence. According to them, the style of reasoning promoted by science, shaped by the scientific method, is the style that makes the best contribution to knowledge. But historicists, whose analysis is inspired by the detailed analysis of specific episodes in the history of science and highlights the complexity of the factors that are often found to determine actual theory change, provocatively compare the replacement of a dominant theory in a domain of investigation to a religious conversion. The scientific community is not a collective rational agent weighing up reasons for and against competing theories in an objective way. It is torn between its natural conservative attitude which would encourage scientists to preserve the existing theories, and the pressure coming from the realization that the existing theories might no longer provide a satisfactory fit with the data. In this context, the choice of a theory over its competitors is not always defended on the basis of purely rational arguments. The commitment to the truth or efficacy of the chosen theory is an act of faith of (a portion of) the scientific community in some metaphysical and methodological assumptions, rather than a consequence of the judgment that the chosen theory is superior to its competitors with respect to neutral evidence and objective standards.

To add to the complexity of this debate on scientific change, historicists view the role of empirical evidence in theory change differently. For the historicist, one cannot easily discriminate between rival theories by relying on data alone, as data are never presented in an entirely neutral fashion and can be interpreted as to lend support to one or the other of incompatible theories. The rationalist might agree that data alone are not always sufficient to discriminate between rival theories (recall the Duhem–Quine thesis we discussed in the previous chapter), but insists that there are objective criteria of theory choice that allow us to see change as an instance of progress.

By the end of this chapter you will be able to:

- Identify the relevant factors in a philosophical account of scientific change.
- List and compare possible criteria for theory choice, and be able to assess the claim that some criteria are more important than others.
- Distinguish and assess rationalist and historicist views of change and progress.
- Identify the implications of the incommensurability thesis for science in general, and for the notion of cumulative progress in particular.
- Examine critically multiple interpretations of the claim that science is rational.

5.1. Revolutions

According to Thomas Kuhn's groundbreaking work on scientific revolutions, the process of scientific change can be affected by a number of diverse factors, and by pressures both internal and external to the scientific community. As the rationalist does, Kuhn recognizes that a newly adopted theory is partly endorsed because it is confirmed by relevant data (at times, it gets an advantage on the basis of its success in so-called *crucial* experiments) and that its newly proposed hypotheses need to be able to explain previously recalcitrant phenomena. But confirmation and increasing explanatory power are never sufficient by themselves to explain change: we also need to pay attention to how the scientific community evolves; to which pressures it receives from political or religious authorities, or from society at large; and to the internal hierarchy and organization of the scientific community, including the methodological principles of scientific investigation, the education of new practitioners, the system of rewards and punishments, and the set of problems that the discipline is supposed to tackle and provide solutions for.

5.1.1. Kuhnian revolutions

Kuhn believes that the proper consideration of these factors often makes it harder, if not impossible, to compare two competing theories on objective grounds. Only after having assessed the view that competing theoretical traditions can sometimes be incommensurable, will we be able to establish to what extent scientific progress can be cumulative. Kuhn (1962, 1970) compares a radical shift to the sudden fall of a government, a *coup d'état*. The analogy with politics implicit in the term "revolution" is not casual. Kuhn believes that, at any one time in mature science, a scientific community is dominated by one major theory that is overthrown only when the tension between the theory and the scientific evidence brings about a crisis of confidence in the theory within the scientific community and plausible alternatives to the theory become available.

   In a scientific revolution, the scientific community is swept by a change that is often radical and multi-layered, and as a result an alternative theory gains dominance. According to Kuhn, as we shall see, political and social factors are combined with the lack of empirical success of a theory to bring about scientific change, and change often involves some radical shift in the language – with the introduction of new concepts or the change in the descriptions associated with previously used terminology. During a revolution not only is the dominant theory overthrown, but also metaphysical views, principles of methodology and other aspects of scientific practice are all subject to revision. The term "revolution" suggests that there is a strong discontinuity between the pre- and post-revolutionary period – which means that the term is better suited to describe some, but not all, cases of scientific change.

 Discuss: Do you find the analogy between political revolutions and radical changes in science compelling?

 Exercise: Compare the French Revolution in the XVIII century and the Copernican Revolution and note any relevant disanalogies.

In the *Logic of Scientific Discovery* and *Conjectures and Refutations*, Popper offers a very different picture of scientific change. Change is dictated by the conditions under which a theory is refuted. When it is shown that predictions made on the basis of the theory conflict with the available data, scientists need to look elsewhere and adopt an alternative theory which has at least the same amount of empirical content as the falsified theory but has not yet been falsified. In this picture, the theory changes, but not much else. Both theories, the new and the old, provide answers to the same problems and are tested in accordance with the same criteria. To develop the analogy with the political sphere, scientific change as described by Popper is not a revolution, but the formation of a new government that has been regularly elected through a democratic process and is going to abide by the same constitution.

5.1.2.   *The Rationalists*

Popper distinguishes three requirements for the growth of knowledge. A new theory must: (a) proceed from a simple and powerful idea that connects facts that were previously unrelated; (b) be independently testable; and (c) pass new and severe tests. The first two are formal requirements of originality and testability. As to the last requirement, the theory to be accepted needs to be genuinely novel and not only explain the phenomena that it is designed to explain, but also lead to the prediction of other phenomena.

Progress, as the succession of theories that are increasingly closer to the truth, is seen by Popper (1975) as a way in which the human species adapts to its environment. When we adopt a new theory, we do so because it solves some of the problems we haven't been able to solve by applying the previous theory. But the adoption of a new theory creates new problems which need to be tackled, and we do tackle them by testing the theory further in the attempt to eliminate errors. This implies that there are both conservative and revolutionary elements in theory change: the new theory is revolutionary in that it needs to conflict with the predecessor in some significant way (and help solve previously unsolved problems); but it is also conservative as it needs to explain why its predecessor did work (at least to some extent). Progress is cumulative in this picture: the new theory has to be seen as an improvement on its predecessor, and therefore comparison between the two theories must be possible.

 Exercise: Can you infer from this brief description of Popper's approach whether he views progress as a rational process?

Popper wants to distinguish the rationality of the scientist making a discovery or testing a theory from the rationality of scientific progress. The agent can make a choice on the basis of an intuition that cannot be rationalized on the basis of the methodological rules of scientific practice, but the rationality of progress in science is not thereby compromised. Evidence from the history of specific episodes in science is scarcely relevant, for Popper, to the assessment of the rationality of scientific progress: scientific revolutions can be accompanied by ideological revolutions, when the discovery which is the focus of the attention of the scientific community seems to support or conflict with, say, a way of thinking, a religious dogma, or the vision of the place of humanity within nature. However, clashing ideologies are not an integral part of the scientific revolution itself and do not affect the rationality of the process that the scientific community undergoes. Popper acknowledges the existence of psychological, sociological, and ideological factors that might influence scientific practice, but claims that these factors can and should be kept separate from an analysis of scientific change and should not affect our judgment about the rationality of progress.

Larry Laudan (1987) is more skeptical about the thought that the rationality of scientific progress can be assessed independently of the rationality of choice of individual scientists. He considers whether there is continuity between scientific methodologies through the history of science and whether these elements of continuity or discontinuity are consistent with the view that change in science operates according to rational principles. Can we construct a coherent description of scientific methodology in terms of the principles of rational agency followed by the scientists who greatly contributed to the progress of their disciplines?

Laudan argues that the project of coming to a *unified* notion of *the* scientific method on the basis of historical evidence is misconceived because any judgment of rationality involves careful consideration of the nature of the actions that were taken, the aims and intentions of the agent in acting in that way, and the background beliefs of the agent about the possible consequences of the actions taken. When we evaluate the rationality of the methods adopted by scientists in the past, we find important differences between the agent's aims and background beliefs and ours, because not only has the shared body of knowledge expanded, but relevant beliefs about methodology have also changed, e.g. about which objectives scientific investigation should have and what the best way of achieving those objectives is. Lack of continuity in beliefs about the aims and means of science is not *per se* evidence of irrationality, but it does suggest that scientific change involves more than the replacement of one theory with another and leads to methodological innovations as well.

Consider the debate in psychology about the reliability of introspective reports and their role in a legitimate scientific enterprise. Prior to behaviorism, psychologists were interested in conscious experience and made use of introspective reports as a reliable way to gain access to it (although they disagreed on what counted as an introspective report). With the behaviorist paradigm in place, behavior became the focus of scientific investigation, and observation of behavior replaced the recording of introspective reports as the means to gather evidence about the object of study. Whether introspection is a means to obtain relevant knowledge does depend on what is perceived to be the purpose of scientific psychology. Introspective reports might be a significant source of evidence for descriptive psychology and phenomenological accounts of experience. However, introspection reports taken at face value do not seem to be useful to investigate aspects of cognitive and social psychology, as they are not a reliable guide to the identification of the type of reasons that motivate people to act in a certain way. This is due to the widespread failure of self knowledge manifested in introspective reports and to the common practice of rationalizing one's own thoughts and actions in order to present a coherent self image. Titchener (1912) provides a good discussion of the legitimacy of introspection at the time of a paradigm shift.

 Exercise: Can you think of another example of a methodological shift in the history of a science?

In the end Laudan proposes to adopt a *naturalized* conception of scientific methodology where a strategy is deemed reliable when used to achieve certain goals on the basis of certain background beliefs. The analysis is naturalized, because methodological rules are subject to evidence as much as the theories are and can be revised or modified accordingly. An example of such a rule is: "In order to arrive at the formulation of reliable theories, avoid *ad hoc* modifications of the theories you are considering." The efficacy of this strategy can be tested, on the basis of the consequences of adopting it, and how progressive the practice of science is when it abides by this rule.

Laudan seems to mediate between the positions of Popper and Kuhn. There is a sense in which Kuhn is right; theories are not the only thing that changes when a scientific revolution takes place. On the other hand, historical analyses of scientific change and the observation of shifting methods and aims of scientific inquiry do not speak either against rationality or against progress.

 Discuss: What is the common ground between Popper and Laudan's analyses of progress?

5.2.   Paradigm Shifts

Before we can start describing the process of theory change that Kuhn calls a scientific revolution in more detail, we need to introduce some terminology, and especially define normal science, paradigms, and anomalies. Then, in the next two sections, we shall illustrate the way in which revolutions work by using the example of the chemical revolution.

According to Kuhn, most of scientific practice is characterized by normal science. Normal science is a period in which scientific research within a discipline is aimed at identifying which facts are important and need explaining, at checking whether the facts observed fit the dominant theory, and at further developing the theory by, for instance, extending its explanatory and predictive power to new areas of investigation. In periods of normal science, researchers consolidate the theory and operate conservatively within a paradigm.

The paradigm is a system that does not include only accepted theoretical statements, but also background beliefs (often of a metaphysical or ideological nature), a set of criteria by which scientists evaluate hypotheses (e.g. accuracy, consistency, simplicity etc.), strategies to formulate and test new hypotheses, and models for the solution of problems which have methodological value and are used also in educational settings, that is, to train junior scientists in that discipline.

When the expectations of the scientific community with regard to the fit between the dominant theory and the observed facts are not met, and the theory does not seem to be confirmed by the data, then Kuhn talks about there being anomalies. Finding anomalies does not by itself condemn a theory, but if anomalies accumulate and undermine the confidence that scientists have in the explanatory and predictive powers of the theory, then a period of crisis ensues. During the crisis, the routine of normal science changes and a more critical attitude is adopted towards the dominant theory. Often, such periods of crisis anticipate a revolution.

5.2.1.   The "discovery" of oxygen

Kuhn himself uses the chemical revolution as an example of how scientific change occurs. The chemical revolution is characterized by a new theoretical account of combustion, the rejection of the phlogiston theory and the discovery of oxygen. It is often presented as the contraposition of two scientists, Joseph Priestley (1733–1804), defender of the phlogiston theory, and Antoine-Laurent Lavoisier (1743–94), who first identified the role of oxygen in combustion and respiration.

Lavoisier's first achievements, though on the way to the rejection of the standard explanation of combustion and calcination, were probably not thought by Lavoisier himself to be incompatible with the phlogiston theory. It is a matter of dispute when exactly Lavoisier abandoned the phlogiston theory, but he certainly oscillated at the beginning of his research. A gradual awareness brought Lavoisier to the mature phase of the oxygen theory, which changed the explanation of many phenomena in chemistry.

While the merit of the theory is undoubtedly Lavoisier's, the claim that he "discovered" oxygen is not uncontroversial. On the one hand, he correctly introduced the notion of oxygen to account for the experimental results he obtained, but he was not the first to perform those experiments and to obtain those results. The novelty consisted merely in his interpretation of the results. On the other hand, we would not today accept the definition of oxygen he provided, because chemical research has developed since then. We could say that, depending on our notion of "discovery," oxygen was discovered either before or after Lavoisier, either by those who isolated it first or by those who first defined it as an element in the way we take now to be correct.

What did the phlogiston theory say? Combustibles contain an inflammable principle that they release upon burning. Many similarities between combustion and calcination were found and metal calcination was viewed as nothing but a slow combustion. It was thought that there were three different kinds of body constituents: (1) air, (2) water, and (3) earths. Earths could be of three different kinds: inflammable earth, mercurial earth, and vitreous earth. On combustion inflammable earth was released. This substance was called also *terra pinguis*, which in Latin means "fatty earth" or *phlogiston*, which in Greek was used to mean "the principle of fire."

What were the properties of phlogiston? It escapes from burning bodies with rapid motion and is contained in all combustible bodies and metals, which can be burnt to calces. The burnt product can be restored to the original substance by supplying phlogiston from any material containing it, such as oil, wax, charcoal, or soot. Zinc on heating to redness burns with brilliant flame, hence phlogiston escapes. The white residue is calx of zinc (Calx of zinc + Phlogiston = Zinc). If the residue is heated to redness with charcoal, rich in phlogiston, zinc is restored. If phosphorus is burnt, it produces acid matter (Phosphorus + Acid = Phlogiston). If acid is heated with charcoal, phlogiston is absorbed and phosphorus is reproduced.

In Georg Stahl's (1659–1734) version, the phlogiston theory explained the phenomenon of burning as expulsion of an inflammable substance from the object burnt (e.g. sulphur) and that of calcination as expulsion of phlogiston from metals (e.g. iron). The theory was very powerful and comprehensive, because, by appealing to phlogiston, it could explain some features of respiration: if air is saturated with phlogiston by combustion, respiration becomes difficult, since respiration itself is removal of phlogiston from the body into the air.

But this theory had to face some evident anomalies. First, it could not explain why after burning metals, calces are heavier than the original metal, given that the substance has released phlogiston in the process.

Second, it was not clear why combustion ceases in an enclosed volume of air and why the volume of air is reduced after combustion. Phlogiston theorists attempted to find different solutions to these problems (giving rise to a proliferation of differently adjusted theories).

In his *Opusculum Chymico-Physico-Medicum* (1715), Stahl defended the view that when a substance burns, it loses phlogiston and thus it weighs less after combustion. This was correctly observed in wood, e.g. because the ashes are less heavy than the original piece of wood. However, the conversion of metals into calces by heating caused an increment of weight. This anomaly was explained by supposing that other particles entered in the calx as a result of the heating process. The common observation that combustion, calcination, and respiration cannot occur in absence of air was also taken into consideration by the phlogiston theorists. If it is air that absorbs and carries away phlogiston, when there is no air left, phlogiston can neither be absorbed, nor be emitted. There was also an answer to the question about the reduction of the volume of air after combustion: phlogisticated air takes up less room than ordinary air, and this was coherent with the common idea that phlogiston had a negative weight. This solution, although compatible with the phenomenon of metal combustion, could not make sense of wood combustion: how can the ashes be less heavy than wood, if phlogiston, contained in wood but not in the ashes, has a negative weight?

Lavoisier drew the first relevant conclusions on combustion and calcination in 1772, when he performed already known experiments and reinterpreted their results. He explained the fact that sulphur and phosphorus gain weight when they burn and that at the same time the volume of air is reduced by supposing that during combustion air is absorbed by them (*fixation*). He did not have any particular conjecture on oxygen, but presumably he became persuaded that effervescing calces contain air, because in calcination metals gain weight as well. Though Lavoisier was immediately aware of the importance of his results and conjectures, he was not exactly sure about what was released – was it the entire air or just a part of it? At this stage it seems that he had not yet rejected the phlogiston theory, as in 1773 he did not rule out the possibility that fixed air was combined with phlogiston.

Before Lavoisier, two chemists managed to isolate oxygen, and found out some of its properties, but neither of them fully understood the role it had in the phenomena of combustion and calcination. One of them was Priestley. His experiments are reported in *Experiments and Observations on Different Kinds of Air*, 1774–7. In 1774 he obtained oxygen by heating red calx of mercury with a burning glass and showed that this new kind of air is insoluble in water and supports the combustion of a candle with a vigorous flame. He called this kind of air "pure air," because he found it pleasant to breathe and thought it could be used for therapeutic purposes. Then, he called it "dephlogisticated air," because he supposed it free of phlogiston. If we burn a candle in ordinary air, the time of combustion is limited, because ordinary air already contains phlogiston and cannot absorb much. If the same combustion takes place in pure air, then the candle burns longer, because the new gas, which does not contain phlogiston, can absorb more of it.

Lavoisier was one of the first chemists to adopt a quantitative method in conducting experiments, for instance he made regular use of a scale. This point is important, because we can see that his taking seriously the anomalies of the phlogiston theory depended on two tacit assumptions, that of the indestructibility of matter and the conservation of mass. The quantitative aspects of the experimental results obtained violated these principles. He made suppositions for justifying the results he obtained with phosphorus, whose weight increases after combustion, and in 1774 he repeated Priestley's experiments, after having met Priestley in Paris. As others did before him, he managed to isolate oxygen, which he called at this stage "the entire air itself." But in 1778, he stressed the fact this gas was purer than the one we live in and defined it "the most salubrious and the most pure portion of the air," as Priestley had done. He also recognized it as the real combustible body and was ready to develop a theory of combustion that was incompatible and alternative to Stahl's phlogiston-based theory. Air is composed of nitrogen, called by Lavoisier *mophette*, and pure air. Pure air is what combustibles absorb and calces contain.

In 1780 he listed the main points of his mature theory:

- In any combustion there is a disengagement of the matter of fire or of light.
- A body can burn only in pure air.
- There is destruction or decomposition of pure air, and the increase in weight of the body burnt is exactly equal to the weight of the air destroyed or decomposed.
- The body burnt changes into an acid by addition of the substance that increases its weight.
- Pure air is a compound of the matter of fire or of light with a base.

In combustion the burning body removes the base which it attracts more strongly than does the matter of heat and sets free the combined matter of heat. The phlogiston theory located the matter of fire in the combustible instead of in the pure air. From 1780 Lavoisier called pure air *principe oxygine*, from which the name oxygen derives. It is a Greek word meaning "acid former," as Lavoisier thought that oxygen was the fundamental constituent of all acids (we know now that some acids do not contain oxygen, e.g. HCl). But Lavoisier's oxygen in 1780 is not an element yet, as for contemporary chemistry, but a compound constituted by *principe oxygine* and matter of fire. Only later, in 1789, would he insert oxygen into his table of elements, together with light and *caloric* (= matter of fire and heat). By appeal to the *principe oxygine*, Lavoisier rejected definitively the phlogiston theory and wrote in 1783

(*Reflections on Phlogiston*) that since everything can be explained in chemistry in a satisfactory manner without the aid of phlogiston, it is probable that phlogiston does not exist.

5.2.2. The chemical revolution as an illustration of Kuhn's theory

As we have seen, the phlogiston theory had to face striking anomalies (e.g., combustibles releasing phlogiston and gaining weight) long before an alternative became available. Kuhn says that counter-evidence has a crucial role in so much as it provokes a tension and makes scientists doubt the efficiency of the theory they adhere to, but anomalies count as counter-instances only in specific circumstances. The loss of empirical adequacy has a *psychological* effect: when anomalies increase, scientists distrust the predictive power of the theory they are committed to and start considering available alternatives. There is no strict methodological rule, in Kuhn's account of how science works, that compels scientists to abandon immediately the theory that does not perfectly fit experimental or observational data. The scientist is more likely to question her own competence in conducting the experiment than the efficacy of the model within which she is operating in absence of other reasons for distrusting it.

The initial occurrence of anomalies does not necessarily represent a decisive threat to the theory; on the contrary, it stimulates research and investigation within it. In the case of the phlogiston theory, this is just the case: the "unpredicted" increase of weight of combustibles was explained in different ways by phlogiston theorists and there was a proliferation of slightly different versions of the standard theory. There are always difficulties, Kuhn says, in the paradigm-nature fit and no emerging anomaly is a knockdown argument against a theory by itself. Kuhn is not optimistic about the possibility of falsifying a theory or choosing between theories by means of pure observation or empirical inquiry. Some rival theories appear to be, up to a certain point, empirically equivalent and no experiment seems able to discriminate between them. In the same way, when there is no competition yet for the dominant theory, baffling experimental results may have little effect on the scientific community. Scientists will naturally attempt first to explain anomalies by means of the tools their paradigm already provides.

Exercise: How does this account of the way in which scientists deal with anomalies in Kuhn's model differ from Popper's account of the method of falsificationism?

When an established theory has to account for phenomena that seem to contradict the predictions made, we witness the phenomenon of proliferation. This proliferation represents scientists' response to emerging anomalies: the scientist who trusts her paradigm tries to improve its explanatory coherence by introducing *ad hoc* hypotheses. An example of this phenomenon is Stahl's attempt to introduce natural levity, or negative weight, to explain why the substances that absorb phlogiston are lighter than before. The result of this process is nothing but copious production, not of alternatives, but of remedies to the limits of the theory.

At a further stage, when anomalies have accumulated, scientists become aware of the increasing difficulties in making the paradigm-nature fit smooth. The problem-solving breakdown is a first step towards the formation of their critical attitudes, even though other reasons may cause discontent towards a theory. In a model where scientists' psychological attitude can determine how they react to anomalies, doubting the philosophical assumptions involved in the acceptance of a paradigm can be relevant to theory choice.

There are two important issues here: (1) whether we are justified in using the concept of discovery when we describe the introduction of some useful concepts which refer to theoretical entities unknown before; (2) what is the relevance of these "discoveries" for the occurrence of a revolutionary paradigm shift. Kuhn introduces a distinction between *discoveries* and *inventions*, slightly altering the ordinary use of these words: discoveries are novelties of facts, while inventions are novelties of theories. Anomalies emerge, the theory is explored and adjusted to account for new facts and then the scientific community is ready for a theoretical shift. Kuhn's example of how theoretical and factual discoveries are intertwined is the discovery of oxygen. He suggests that we should ascribe the discovery of oxygen to Priestley and its invention to Lavoisier, as Lavoisier but not Priestley realized the theoretical implications of the discovery of oxygen. This remark indicates what Kuhn thinks about a discovery: first, it is meaningless without a corresponding invention. We could not even say that Priestley discovered oxygen unless Lavoisier's subsequent theory had shown what oxygen is and how combustion works. More generally, no new fact is relevant without being interpreted theoretically. The thought is that a "new" fact by itself cannot be used either to confirm or disconfirm a theory: a discovery has effects on scientific theory-dynamics only when the emergence of a phenomenon that has not been considered before (that is *novel*) is included in a global process of theoretical re-interpretation. In this context, three chemists isolated oxygen, but only one recognized it as a distinct substance that plays a part in combustion and calcination. Priestley first isolated what we call oxygen now, but did not make any conceptual discovery, as he treated it as dephlogisticated air, including it in the explanation provided by the phlogiston theory.

 Exercise: How does Kuhn's distinction between discovery and invention map onto the common use of these terms?

The relation between theory and world cannot be considered separately from the relation between a theory and its rival. In fact, discoveries seem to occur before a paradigm revision and anticipate limited or extended revolutions. This suggests that for scientists it is easier to recognize new facts in the phase in which they doubt the efficacy of the paradigm as a problem-solving device. We saw in our brief description of Lavoisier's gradual approximation to the oxygen theory that at the beginning of his research he was not confident enough to reject the phlogiston theory all at once, as he did not have any alternative theory of combustion. But his reflections about the role of oxygen in combustion and calcination were decisive and brought him to a complete rejection of the standard view.

Kuhn maintains that, for the conservative character of normal science, scientists trust their model and resist anomalies, unless there is a competing model that is better at dealing with at least some unsolved problems. He employs the notion of *mature* times. Revolutions in their occurrences are sudden and dramatic, but they cannot occur without being anticipated by evident symptoms of crisis. A field can be mature or immature for change, and in the initial stages of scientists' detachment from their paradigm it is not already clear whether the paradigmatic theory just needs improvements and adjustments or is bound to be replaced.

Kuhn tends to give a *psychologistic* picture of theory change, implying that often scientists are not aware of the reasons why they make their choice, either to preserve the old paradigm or to embrace a new one. Some philosophers of science (Thagard 1993 and Laudan 1977) criticize Kuhn's account and describe the shift from an established theory to a competing one as a gradual exploration of new possibilities conducted consciously by the scientific community. However, it must be noted that Kuhn does not view the resistance to a new theory as irrational just because it depends on the obstinate adhesion to a paradigm within which scientists have been trained and have been working. After all, constancy in defending what is taken to be an established truth is recognized as a rational virtue. Kuhn concedes that something is amiss in Priestley's refusal to embrace the oxygen theory when he resists Lavoisier's arguments and remains isolated in the scientific community by continuing to accept the phlogiston theory. Priestley, Kuhn says, ceased to be a scientist, when he isolated himself in order to continue his work, because he ceased to engage with a community of practitioners. Were we to consider his behavior as irrational, we would not then attribute irrationality to science in general.

> Discuss: Do you agree with Kuhn that one cannot be a scientist in isolation from a community of practitioners with shared knowledge and common goals? Can there be science in a world populated by only one individual?

When Kuhn describes revolutions as changes of worldview he has in mind a particular thesis, that scientists after a revolution see (and work in) a different world. This seems the most evident denial of realism among Kuhn's many striking claims, but realism is not at issue here. Kuhn never denies that Lavoisier and Priestley were *looking at* the same chemical world, at the same gases and combustibles. He denies that they were *seeing* the same objects. Considering the ontological aspect of a paradigm, we understand Kuhn's stress on the word "see" rather than "look" or "interpret." Scientists committed to rival paradigms may look at the same world, but see different things. Coming back to Lavoisier and Priestley, they both "looked at" pure air; Lavoisier saw oxygen, while Priestley saw dephlogisticated air. This implies a difference of world-views and not of worlds. They did not "interpret" what they saw in different ways, but they saw different things. If we spoke of interpretation, we would assume there is a shared *datum* to interpret, a *given* to classify in the light of different conceptual schemes and categories.

But Kuhn argues that there is no given in science, that data are always collected with difficulty and that they are not neutral or pure when they become accessible to scientists, because they become an integral part of a system of background beliefs and metaphysical assumptions, methodological guidelines and notions learned during their training. These beliefs shape the scientists' knowledge and their way of seeing the world, so that, when the community witnesses a paradigm shift, the world of their experience changes.

 Exercise: Choose another important theoretical shift in science (e.g. from Newton's physics to Einstein's relativity; from behaviorism to cognitivism in psychology; Darwin's revolution in biology) and do some research on it. Then assess to what extent that particular shift fits Kuhn's account of scientific revolutions.

Discuss: Is there a connection between the thought that the world changes after a paradigm shift and incommensurability?

5.3.   Beyond Revolutions

The chemical revolution seems to be a perfect illustration of Kuhn's model of theory change and of the various phases of a revolution. But other philosophers have found Kuhn's analysis of scientific change either misleading or applicable only to a very limited set of cases in the history of science. What are the alternatives to Kuhn's account? We shall review some of them in this section.

5.3.1.   Research programs

Among Kuhn's many critics, Imre Lakatos is particularly noteworthy, as he attempts to mediate between the novelty of Kuhn's analysis and

Popper's more conservative views about progress and truth. Lakatos appreciates Kuhn's recognition of the importance of the role of history and case studies for the specification of scientific method. He also concedes that it is right to introduce sociological, psychological, and political factors (what the positivists took to be "external factors") in the explanation of scientists' behavior and to reject a naïve version of falsificationism. On the other hand, Lakatos criticizes the vagueness of the notion of paradigm (which makes it theoretically confusing) and its metaphysical connotations as a worldview. He also denies the plausibility of presenting revolutions as conversions, and the incommensurability thesis. His purpose is to promote a rational reconstruction of science by redefining scientific progress and providing a more sophisticated version of falsificationism than Popper's.

According to Lakatos, Kuhn describes science as irrational, because, in discussing the nature of change, he speaks of revolutions as religious conversions and this, according to Lakatos, amounts to viewing scientific change as irrational blurring any demarcation criteria between science and non-science. What might Kuhn have implied by his analogy between scientific change and conversions? Conversion is a change not only of a set of beliefs, but also of many fundamental assumptions on which those beliefs were grounded. Kuhn is not the first author to compare intellectual change of beliefs to the experience of conversion. In *On Certainty* the philosopher Ludwig Wittgenstein (1889–1951) explores the problem of the relation between skepticism and commonsense and focuses his attention on a special kind of propositions (so-called *framework propositions*) that we hold for certain and rarely revise. These propositions look like empirical propositions about the world, but their role is almost grammatical, as they describe our way of seeing the world rather than the world itself (e.g. "The world had existed long before my birth," "I have two hands"). If we were to reject these propositions, we would embrace a totally different worldview. And if I were to persuade the inhabitant of a remote and culturally isolated tribe who was taught and believed otherwise that the world came into existence before he was born, it would be difficult to persuade him by rational argument alone. I could tell him that I have learned in history, a subject I was taught at school, that people lived, prospered, built monuments and fought in wars long before we were born. I could tell him that I know people, among which are the people I call my parents, who are older than me and have lived longer than I have. But these considerations are not *independent* arguments for the view that the world existed long before our births, as we would not take them seriously at all unless we already assumed that the world had existed long before our birth. If the people in the tribe were to switch to my worldview, they would not do so under the pressure of rational argumentation, because all arguments in favor of the proposition that the world did not come into existence when we were born somehow already presuppose the truth of the proposition.

Exercise: Can you imagine a situation in which a framework proposition is revised or rejected?

 Discuss: Can the case for conversion made with respect to framework propositions be made with respect to paradigm shifts as well?

Lakatos recognizes that there are some aspects of a research program which are less likely to be revised (even in the face of counter-evidence), and he calls those the *core* of the research program (e.g. basic tenets of a theory). Other aspects are more flexible and less strenuously defended and he calls them the *protective belt* of the research program (e.g. auxiliary hypotheses). When disconfirming evidence becomes available, the belt is doubted before the core, and different strategies can be adopted to "correct" the research program. This account is preferable to that of Kuhn, Lakatos believes, because it is historically more realistic and does not compromise the rationality of change. For Lakatos, the reasons for changing the protective belt or preserving the hard core can be either good or bad reasons depending on whether the adjustments are *ad hoc* or whether they lead to novel predictions.

If Kuhnian paradigms are like worldviews, then some of their constituents are such that we cannot abandon them without changing paradigm. There may be no inter-paradigmatic good reasons for the change, as the change is so radical that what counts as a good reason for it is also subject to change. However, this does not necessarily mean that the change occurs without reasons or for bad reasons. After a scientific community is converted, then one can find reasons to explain the shift and generally these can be found in the fact that a new paradigm accounts for areas of scientific investigation that were not satisfactorily explained before the shift. Lakatos argues against this notion of rationality as *post-hoc* rationalization, and leaves no room for it in his description of the way in which one research program replaces another. The main difference between paradigms and research programs is that the validity of the latter can be assessed objectively, from a point of view that need not be internal to the research program under consideration. As anticipated, research programs are sequences of theories within a certain domain and are constituted by a theoretical core and auxiliary hypotheses (the protective belt). When the predictions made in accordance with one research program are falsified it is not always clear what should be rejected (recall the Duhem–Quine thesis and its challenges for falsificationism). The activity within a research program is guided by methodological heuristics which help scientists decide what is most likely to be responsible for a failed prediction, the theoretical core, or one of the auxiliary hypotheses. If a theory needs to be rejected and replaced, a new problem within a research program will take center-stage and the shift can be either progressive or degenerating. In order for the shift to contribute to progress, the new theory needs to be more informative than the previous one; needs to be able to explain the success of the previous one; and needs to receive independent corroboration.

Lakatos suggests that Kuhn's picture of the history of science as a series of periods of normal science interrupted by a revolution is inaccurate: only rarely in science was there genuine monopoly. The proliferation of alternative solutions to existing problems is not only a feature of periods of crisis, according to Lakatos, but it is an essential element of scientific practice: the choice between competing theories within research programs, and between research programs, is required all the time, not just during a major revolution, and making choices is the only way to generate progress. If paradigms are regarded as all-comprehensive worldviews, in the sense that once we are committed to a new paradigm we see the world in a different way and come to accept partially new standards for theory choice, then the proliferation of competing paradigms does become an exceptional situation. But we should reject, according to Lakatos, such a conception of paradigms.

Lakatos attacks in particular the idea that each paradigm has its own rationality. Serious problems arise when we ask ourselves to what extent paradigms are independent and cannot be objectively evaluated, because Kuhn's position leads to the theses of incommensurability. "Incommensurability" can be used in different ways – it refers to the impossibility of comparing statements from rival theories, but also to the lack of common methodological rules on the basis of which the conduct of scientists belonging to different paradigms can be evaluated; and to the lack of common criteria for theory choice, independent of a paradigm. All of these incommensurability theses are challenged by Lakatos, according to whom methodological strategies and criteria for theory-choice are not theory-dependent.

Discuss: Is the account of theory change that Lakatos provides a genuine improvement over the Kuhnian notion of revolution? Can Lakatos save the notion of rationality and progress within science?

5.3.2.   Styles of reasoning

In the *Structure of Scientific Revolutions*, Kuhn argues that competing paradigms are actually incommensurable and uses the provoking conversion analogy: the shift from one paradigm to another cannot be entirely due to arguments, because what counts as an argument supported by good reasons for scientists adhering to the former paradigm can be ignored or count as a bad argument for scientists adhering to the latter. Paradigms do not just have languages that are resistant to interparadigmatic translation, but come with sets of values that are used in assessing whether arguments are good and theories are successful. If these criteria are internal to a paradigm, no choice between paradigms can be justified by appealing to those values, unless they are shared by the paradigms that we are interested in comparing. Different paradigms can redefine the aims of science, and propose different conceptions of evidence, confirmation, and explanation.

An alternative to the notions of paradigms and research programs is that of styles of reasoning, or traditions. Both Hacking (1982) and Feyerabend (1975) use this terminology, but arrive at different conclusions about the way in which styles of reasoning determine answers to ideological and methodological questions and how they relate to one another in periods of both proliferation and change. Fundamentally, the problem that both Hacking and Feyerabend approach, in different ways, is whether there are elements of relativism in any historically accurate account of scientific and conceptual change. In particular, they are not interested in the body of true beliefs that can be maintained or revised any time a new theory is put forward, but in the way of thinking which is informing our investigation of nature, and the criteria for what counts as a good reason to support a belief change.

Feyerabend starts from a comprehensive critique of any account of scientific methodology that prescribes unchanging rules on the basis of the observation that in the history of science different strategies have been successful. These strategies cannot be seen as exceptions to a set of rules: the fact that they were adopted and delivered results is rather an indication that the proliferation of methodological approaches is desirable. He continues to argue that for any rule codified by philosophers of science there is an opposite rule that is just as acceptable, and claims that in the history of science bigger steps have been taken when scientists have been less conservative and have violated explicit methodological rules such as consistency by, for instance, putting forward hypotheses that were not compatible with the theories that were regarded as established truths at the time.

Discuss: Does the observation of a plurality of methods in science support the conclusion that there is no way of codifying scientific methodology? Does it support the view that there is no rationality in science?

Within science, then, there is a plurality of incompatible and equally acceptable methods. For Feyerabend it is important to highlight that thoughts that were not arrived at in a legitimate scientific way have contributed to the direction in which science has progressed. Even ideas that were imposed by prejudice and other "irrational" tendencies have been essential to the process by which some of our current theories have won over their competitors. Given his commitment to recognize elements conducive to progress in different traditions of thought, Feyerabend is skeptical about the way in which scientists market their research programs as guided by rational principles and finds them at risk of limiting future progress by imposing constraints on the sources of ideas that are regarded as acceptable. To reject an idea only because it seems to have been generated by the adoption of a long-forgotten cosmology, for instance, would be a mistake and a myopic choice. The thesis is that methodology varies according to the tradition of thought within which an idea emerges or an inquiry is conducted; and that even within science, which we are tempted to see as a methodologically unified enterprise, almost anything goes.

Exercise: Think of an example of a discarded viewpoint that has generated what we would now consider as legitimate scientific hypotheses.

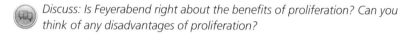

Discuss: Is Feyerabend right about the benefits of proliferation? Can you think of any disadvantages of proliferation?

Hacking shares with Feyerabend an interest not in theory change *per se*, but in the variation noticeable in the ways in which we justify those theories, in the style of reasoning to which we explicitly or implicitly subscribe when we regard a problem as a legitimate problem. It is not surprising that some people find true what others find false. This type of disagreement can persist within a tradition of thought, without generating any substantial clash or revolution. What seems to be a more radical form of change is the shift in styles of reasoning: Hacking uses the example of the alchemical and astrological doctrines of resemblance and similitude in the Renaissance (1982, p. 60). We do not understand these explanations as science because we now employ different concepts (e.g. the concept of evidence has changed) and have completely different reasons for believing that a hypothesis is explanatory (e.g. we can understand the claim that mercury salve is good for syphilis but we fail to understand the reason for it, that mercury is linked to the planet Mercury which is linked to the marketplace, where the disease is contracted).

Bird (1998) offers another good example of a type of explanation that Hempel regarded as un-scientific: the astronomer Francesco Sizi argued that the number of the planets is necessarily seven because we have got seven "windows in the head" (two nostrils, two ears, two eyes and the mouth) and there are seven metals. As Bird observes, the symmetry could be defended by appealing to the intentions of a creator and would have counted as a reason to believe that the number of planets is seven in that context. But in contemporary science, the way Sizi argues for the necessity of the existence of seven planets is not acceptable. These examples show that when the style of reasoning changes, the type of explanation and justification that is considered as acceptable changes and, according to Hacking, understanding can break down for this reason.

Hacking disagrees with Kuhn about there being impossibility of translation between the statements of a theory and the statements of its rival. But he agrees that there is discontinuity in conceptual change, where the gap comes from a variation in the way of thinking, and in particular in what is considered to be a good reason for supporting a claim. Because it is reasoning and methods which are relative to a particular style, styles cannot be successfully compared to one another and judgments of superiority cannot be independently justified.

Discuss: Do you agree with Hacking that understanding breaks down across styles of reasoning? Can you think of other examples of styles of reasoning that were regarded as scientific at one time but would be rejected as un-scientific now?

5.3.3. Theory choice

Whether we believe that paradigms, research programs or styles of reasoning are the most suitable units of scientific change, we encounter serious difficulties in comparing rival theories when the change in the description of the theoretical terms and conflicting methodological assumptions raise the issue of incommensurability. The incommensurability of meaning and methods has implications for scientific progress and theory choice. If statements of rival theories cannot be compared for lack of a common language, of common concepts, then it becomes extremely difficult to choose between them on the basis of their empirical adequacy. How can we argue that one theory has *more* empirical content than the other, or that it performs better the tasks of predicting and explaining the phenomena in the same domain? The idea that new theories can account for the partial success of their predecessors seems to be a fundamental step in establishing a notion of progress which is genuinely cumulative.

But of course the claim that rival theories cannot be compared on the basis of their content, and therefore on the basis of their empirical adequacy, does not rule out that there are other ways to distinguish between them and make a rational choice. Theories can be compared with respect to criteria that can provide us with reasons for justifying that choice. What are these standards by which we evaluate theories? In the postscript to the *Structure of Scientific Revolutions*, Kuhn lists some of these values: a theory should make accurate predictions (possibly of a quantitative and not just qualitative nature); it should allow problem-solving; it should be simple and consistent; and it should be largely compatible with other accepted theories. Newton-Smith (1981) lists other values: a theory should be fertile, and allow for theoretical and practical developments; and it should be well-integrated not just with other accepted theories, but also with some general metaphysical assumptions about the world.

The lists are very heterogeneous. Some criteria seem to be purely a matter of aesthetical judgment (e.g. the elegance of a mathematical proof). Other criteria seem to track important epistemic virtues: as we remarked in chapter 1, the good integration with other accepted theories is a reason to prefer psychology to astrology when looking for a theory that can provide an account of human behavior. But arguably all the criteria track epistemic virtues: the extensive debate on simplicity or parsimony as a *desideratum* for theories shows that it is not easy to discriminate in advance of empirical investigation the possible implications of criteria for theory choice.

 Exercise: Can you add some other criteria to the list?

Kuhn argues that the values are fairly constant across paradigms, but that the application of these values can be subject to individual differences, and differences between scientific communities: e.g. to what extent a new theory can be in tension with another established theory before the tension becomes an obstacle to the acceptance of the new theory is relative to the judgment made in a particular context, and other evaluations and interests. Another judgment that might depend upon the methodological assumptions in a scientific community is the relative importance of the criteria and whether all criteria are applicable to all scientific theories.

 Discuss: Would it be rational to choose a simpler but less fertile theory?

Summary
In this chapter we have considered different philosophical approaches to change in science and their consequences for the claim that science is a rational enterprise. According to both rationalists and historicists, it seems that the rationality of science is hostage to scientific practice being untouched by those factors that have no direct relevance to the confirmation or disconfirmation of theories. Rationalists believe that theory change is rational and that there is cumulative progress because they think they can reconstruct the process of change in a way that does not need to take into account factors external to the conditions for testing and the method of science. Historicists often deny the rationality of theory change and the cumulative nature of progress because they do not think that the process of change can be reconstructed independently of ideological, social, psychological, political, and historical factors.

But we should challenge the shared assumption in this debate: why cannot there be rationality in change after we recognize that change is not just affected by the objective merits of a theory and its fit with available evidence, but also by the way in which the scientific community as a whole operates? It is only by viewing science as part of society that the import of its contribution to the body of knowledge shared in our communities can be fully appreciated. It is because it has at times resisted, and at other times succumbed, to pressures from "outside" that science has become what it is now, a social institution rather than a mere set of academic disciplines united by some abstract method.

Preview of future attractions

We have started seeing some of the mutual interactions between science and non-science: metaphysical assumptions, ideological and conceptual shifts, psychological features of human motivation on reasoning and theory choice all contribute to shape the trajectory of scientific progress. In the next chapter we shall focus on the ethical responsibilities of science towards the rest of society.

Issues to think about

1. What notion of scientific progress is compatible with Kuhn's account of theory change?
2. How do research programs compare to paradigms?
3. Do you think it makes sense to talk about a theory-neutral "given" in science?
4. Can there be genuinely cumulative progress?
5. What criteria are legitimate criteria for theory choice?
6. Is it possible (and if so, desirable) to postulate a notion of rationality independently of the requirements of scientific methodology?

Further resources

In this chapter I have offered an illustration of Kuhn's views by reference to the chemical revolution. You could compare his account with other accounts of scientific change on the basis of some other examples of significant changes in science, such as the Copernican revolution (Henry 1997; Hall 1983; Kuhn 1957, 1990; Cohen 1980). Other options include the shift from Newtonian physics to relativity theory, or the acceptance of Darwin's theory of evolution. Additional readings on Kuhn can help you understand the significance of his contribution and the details of his proposal: for instance, see Bird (2000) and Hoyningen-Huene (1993). Kuhn's ideas about the structure of revolutions in science have been applied to cognition and computation in the work of Andersen et al. (2006) and Thagard (1992) which focus on concept acquisition and revision.

Useful articles on the coherence and plausibility of the notion of "scientific revolution" can be found in a collection edited by Hacking (1981). Alternative views to Kuhn's theory of revolution and progress can be found in the writings of Popper (1963, 2002, ch. 10), Laudan (1984), and Lakatos (1970). Collections of papers on scientific change and the nature of progress can also offer you an idea of the structure of the debate: see for instance, Radnitzky and Andersson (1978); Niiniluoto and Tuomela (1979); Lakatos and Musgrave (1970); Harré (1975). For a general introduction, see Losee (2003). For more recent contributions to the debate on scientific progress from philosophy and the social sciences, see Bird (2007), Chang (2007), and Lohmann (2004).

6 Ethics

The historicist conception of theory change leads us to consider the interplay between scientific communities and society at large with respect to shared ideologies, conflicts of interest, and financial pressures. One respect in which science and the rest of society often collide is the understanding of the way in which ethical constraints should be placed on the development and funding of research projects and on the technological advances that result from them. Another respect worthy of philosophical investigation is whether there are any obligations towards science, that is, whether supporting science is a moral imperative for individuals and societies.

In this chapter we shall examine the relation between science and the rest of society and the issue of whether scientists are accountable in a way that is different from the way in which other individuals in society are. We shall consider some concrete cases, including the ethics of enhancements and the use of deception in social psychology research, but many more issues could be taken as illustrations of the current debates on the ethical consequences of the methods and objectives of scientific research. Here are some examples: the ethics of stem cell research on early human embryos; the ethics of biomedical research involving non-human primates; issues raised by global warming; the ethics of clinical trials in developing countries; and the ethics of epidemiology research in psychiatry.

The cases we shall be looking at have the function to illustrate the interplay between science and ethics and suggest ways in which scientific research could be constrained by basic principles such as the autonomy of the individual, the avoidance of unnecessary suffering to sentient beings and the obligations we have towards future generations. Out of the discussion of the chosen examples a general view of the relation between science and the rest of society will be developed and the role of political and religious authorities, scientific experts, and the general public in debates involving science will be questioned.

What we should critically examine is the popular assumption that the interests and values of individuals and societies need to be safeguarded from the blind ambition of individual scientists attempting to "play God," from "Frankenstein science," or from some evil corporate interests that hide behind the search for truth and progress. Although there are concrete risks (ethical and otherwise) in the pursuit of scientific research, the rhetoric with which science is often depicted in movies, novels, and even the popular press is often unjustifiably negative. This is arguably due to the failure of society to create opportunities for public engagement with science, where concrete problems that affect scientists and non-scientists alike are discussed in an accessible and transparent way.

By the end of the chapter you will be able to:

- Analyze present ethical debates about the methods and objectives of scientific research.
- Identify the ways in which interests in science might conflict with or promote the interests of individuals in society.
- Account for the complexity of the relation between science and society in terms of mutual obligations and responsibilities.
- Form an opinion about the extent to which scientific research should be autonomous and whether the principle of freedom of research can be justified.

6.1. Instrumentalization

There are many reasons why scientific research as we conceive it today could be seen as raising ethical issues or needing ethical regulation. For instance, one might ask whether it is legitimate to conduct experiments on humans or non-human animals when some health risks are present, or when pain and suffering need to be inflicted as part of the experimental design.

There is one argument for the claim that in research activities involving human or animal subjects, it is morally objectionable to use human beings or animals exclusively *as a means*, e.g. in order to gain knowledge, when these individuals do not stand to benefit themselves from the outcomes of the research. This argument does not depend for its persuasiveness on the evaluation of potential risks. Even if the risks research subjects are subject to is negligible, one could argue that the use of individuals with rights and interests for purposes that might not directly benefit them is morally dubious.

The argument is often traced back to Kant, who believed that we should never use our fellow humans as means only (Kant 1785). In recent years, the view has been extended by animal rights theorists to the protection of some animals. For instance, Tom Regan (1983) has argued that mammals aged one year or more should never be used in research because they have certain capacities that ground the ascription of interests to them, and a weak version of autonomy. According to this line of thought, various activities that can lead to (a morally wrongful form of) instrumentalization, i.e. the use of another individual solely as a means for accomplishing one's own ends, should be avoided even if the actual risk of humans or animals being harmed is low. For utilitarians, instead, there is no straight-forward answer to the question whether in general humans or non-human animals should be used in research (Singer 1974). For each case, one needs to calculate benefits and risks and decide what the right course of action is, depending on factors such as: (1) the significance of the expected experimental results; (2) the amount of pain, stress or suffering that is going to be inflicted on the participants; (3) the level of complexity of the psychological capacities of the individuals involved; (4) the existence of alternative methods of investigation that are reliable; and so on. An experiment with great potential benefits which involves a small number of rats but does not cause them significant pain, might be acceptable. An experiment which is less promising in terms of the significance of the expected results and would involve both confining and inflicting pain to a large number of primates, might not be acceptable.

 Exercise: Can you provide some examples of instrumentalization?

 Discuss: Is using other people or animals for one's own purposes always morally objectionable?

 Discuss: Do you think that it is always morally permissible to use non-human animals in biomedical research?

The notion of instrumentalization needs unpacking. In scientific research, using human participants might not be morally problematic, if certain ethical principles, such as the respect for personal autonomy, are taken into account. Often this principle requires that informed consent is obtained from the participants, that is, that they are openly informed of the risks involved in their participation in the research and the significance of the study is clearly explained to them. If it is true that scientific research benefits all humans, whether it be pure or applied, then participants themselves might have an interest in research activities to be pursued and developed, as long as the risks to their own physical and psychological health are low. One could argue that "using others" is not a prerogative of scientific research. Many of the human activities commonly accepted and sometimes even promoted in our societies (e.g. politics, commercial enterprises, advertising) and many forms of social interaction between individuals or groups of individuals (e.g. friendship, marriage) involve some form of instrumentalization. Instrumentalization is not necessarily a morally objectionable aspect of our human practices, but becomes morally objectionable when it is a case of exploitation, that is, when the interests of our fellow humans or other animals are not respected in the use that we make of them (e.g. slavery). Although exploitation of other humans is widely recognized as morally impermissible in our society, the issue of the exploitation of other animals is much more controversial with respect to common practices such as intensive farming or medical research. There is no consensus on whether non-humans have moral status, and even when it is conceded that they do, it is not clear what having moral status implies for their treatment.

In ethics philosophers often implicitly or explicitly correlate what an individual is entitled to from a moral point of view with the complexity of the mental life of that individual. This correlation (the so-called "psychological approach to moral status") gains center-stage in many attempts to answer the question whether we should accord rights or moral status to those individuals that lack the capacities persons usually share, such as the capacity for rational deliberation and self-consciousness.

Exercise: There are other approaches to establishing which moral obligations we have and to whom. For instance, one view is that special protection should be accorded to the most vulnerable whether or not they have comparable psychological capacities to ours. Can you think of an example in which this line of reasoning can be applied?

The debate on the moral permissibility of scientific research involving non-human animals and human embryos (e.g. cancer research on mice or some form of stem cell research) is a good example of the way in which public opinion and society at large impact on science, and especially on the way in which some research objectives are pursued. Nobody disputes that advancing medicine and discovering treatments for conditions as debilitating as Alzheimer's are ethically legitimate research aims, but the issue is whether any means of achieving those aims is permissible, or whether some constraints should apply to which beings can be used, made to suffer and destroyed for these purposes. Apart from the issue of instrumentalization and exploitation, and other ethical constraints that apply to research *methods*, there is also a lively debate on which *aims and objectives* publicly funded research projects should prioritize given the limited resources allocated to science, and whether some research objectives should be ruled out altogether for ethical reasons.

We are going to explore these issues by looking at some specific examples and attempting to draw some general conclusions about the shape that these debates can take and the way in which progress can be made with respect to them.

6.2. Ethical Constraints on Research Objectives

There are reasons why the objectives of a research proposal could come under ethical scrutiny for two quite independent reasons. There can be moral arguments to the effect that public resources for research are not unlimited and should be distributed fairly, by prioritizing research areas in which greater benefits can be obtained for e.g. the greater number of people or the least well-off – depending on the underlying concept of justice. This kind of limitation on the objectives of research is controversial, as it might subject the progress of science to political decision-making on resource allocation issues.

There can also be reasons to stop a research proposal which has as its aim the demonstration of a thesis that is ethically dubious, as for instance the superiority of one race over another. Many research programs active in the period in which Germany was taken over by Nazism had as an explicit or implicit objective to establish the inferiority of mind of the Jewish population. One instance of this attempt at "racial psychology" was the gathering of evidence for the claim that some intellectual traits necessary for accomplishments in mathematics were racially distributed: the hypothesis was that Jewish mathematicians were good at employing analytical skills (like other Latin populations) and German mathematicians had more developed qualities of imagination and intuitions. These very claims – completely unsupported by reliable evidence and based on anecdotal information – developed into objections to particular styles of doing mathematics (see Ludwig Bierbach's articles on the vulgarity of Jewish mathematics in the 1930s) and were used as a reason to boycott lectures by Jewish colleagues or deny them academic appointments.

More recently, research in biological or chemical warfare may serve as an example of scientific research with morally objectionable objectives, as the potential applications of the research are likely to be used to bring harm rather than to promote well-being.

 Exercise: Can you think of other research projects whose main aim could be regarded as objectionable on moral grounds?

 Discuss: Do you think that moral concerns should ever act as a constraint on the acceptability of a research project given its aims?

In the next section, we are going to examine an example of a research project whose aim (genetic enhancement) has been regarded by some ethicists as morally unacceptable, and by others as something we should be morally obliged to promote.

6.2.1.   Two notions of disability

Suppose that you are soon to be a parent and you learn that there are some simple measures that you can take to make sure that your child will be healthy. In particular, suppose that by following the doctor's advice, you can prevent your child from having a disability, you can make your child immune from a number of dangerous diseases and you can even enhance her future intelligence. All that is required for this to happen is that you (or your partner) comply with lifestyle and dietary requirements. Do you and your partner have any moral reasons (or moral obligations) to follow the doctor's advice? Would it make a difference if, instead of following some simple dietary requirements, you consented to genetic engineering to make sure that your child was free from disabilities, healthy and with above-average intelligence?

In the debate on the ethics of genetic enhancement, one particular move is to argue that, if we agree that disability should be avoided, then we should also agree that enhancements should be pursued, as disability and enhancement seem to be on a *continuum*. Of course, this view depends on the chosen account of disability and relies on the harmed-condition account of disability (Harris 1992), which is to be contrasted with the social conception of disability (Reindal 2000; Koch 2001).

In the harmed-condition account, it makes sense to claim that certain conditions of the person can be regarded as disabling given the physical or social environment in which the person is embedded when they are harmful conditions that a rational person would have a preference not to be in. What all disabling conditions have in common is that, to some extent, they cause harm to the people who are in them (by exposing them to risks, impairing them in what they do, limiting their opportunities or preventing them from having experiences that are worthwhile). On this view, disabling conditions constitute a disadvantage with respect to relevant alternatives, not necessarily with respect to the conditions of the typical human. Changing environmental factors, or new discoveries about the onset of serious diseases, for instance, might make it the case that typical conditions of our species become disabling.

According to the social conception of disability, all disabling features of the condition would disappear if society were inclusive and free from discrimination or prejudice. Whereas it is certainly true that certain attitudes in society towards people who are perceived as different cannot but make things worse for disabled people, in many cases, perhaps most cases, their harmful condition would persist once society had been reformed (e.g. deafness or Down's syndrome). The harmed-condition account is generally preferred to the social conception of disability, because it can explain how certain conditions of the person remain disabling even after issues of discrimination are addressed and society is freed from prejudice.

 Exercise: How could a defender of the social conception of disability reply to this argument?

A case in which a disabling condition would cease to be harmful if society were reformed is the case of a disability entirely caused by the adverse social context. For instance, history teaches us that being born female in nineteenth-century Europe would have been a disabling condition when compared to being born male. Exceptions apart, women were prevented from exercising any form of autonomous decision-making and often lacked the opportunity to have an education. Similarly, being born today in a developing country rather than in a developed country is, for most, a disabling condition because of the consequences for health provision, education, job opportunities, and many other aspects of one's life.

These harmful conditions would cease to be regarded as disabling if we could change the social, political, and economic context. The harmed-condition conception of disability can recognize that disabling conditions might have a variety of causes without being committed to the view that by changing social, political, and economic factors all disabilities would disappear.

From the perspective of a harmed-condition account of disability, one can explore the relation between disability and enhancement. The harm–benefit *continuum* is the idea that there is continuity between the reasons for not harming others and the reasons for benefiting them. It would appear that, if we have moral reasons to prevent people from being in disabling conditions, we might also have moral reasons to improve their conditions, whether they are disabling or not. Further questions are whether these moral reasons are (in at least some cases) moral obligations and whether it is wrong not to enhance individuals whom we could thus benefit. We shall look at some common objections to enhancement and conclude that there are at least three possible ways of conceiving the ethics of enhancement that are compatible with the harmed-condition account of disability.

It would seem that we have moral reasons to prevent disabling conditions, as part of our commitment to the basic moral principle of avoiding unnecessary harm. This means that, when we have the choice, we should bring into existence people without (known) disabling conditions rather than people with such conditions. But this of course says nothing about whether these moral reasons give rise to a moral obligation or about the way in which we should carry out such a moral obligation. The following list includes only some of the ways in which creating a person with a disability can be avoided: postponement of conception, behavior modification, gene therapy, selection between pre-implantation embryos and abortion. One might recognize the moral obligation to prevent or eradicate disabilities and still object on moral or other grounds to the methods by which the obligation can be carried out. Moreover, the strength of the obligation might vary in accordance with the context of the disabling condition, and in accordance with the degree of harm that the disabling condition is likely to cause to future persons.

Exercise: Consider the following example (from Harris 2004) and identify the ethical reasons that would be relevant to making a decision in this case. A woman has six pre-implantation embryos in-vitro awaiting implantation. Three will develop asthma and three seem healthy. Which should she implant? Now consider the following case and explain how it differs from the previous one: A woman is told that if she conceives immediately she will have a child with asthma but that if she postpones pregnancy, takes a course of treatment and then she conceives, she will have a healthy child. What should she do?

6.2.2. Objections to enhancement

If we accept that there are moral reasons to prevent or eradicate disability when possible, does this commit us to recognize that we also have moral reasons to enhance? This question arises with particular force for those who believe that there is a *continuum* between harms and benefits. On this view the reasons we have to avoid harming others are continuous with the reasons we have for conferring benefits on others as we can. On some theories of responsibility for action, where we can protect people from harms, to elect not to do so is to become responsible for the harms we might have prevented. All actions are re-describable as omissions and vice versa. The decision to save a life is the decision not to allow someone to die.

 Exercise: Can you think of counter-intuitive implications for this view of moral responsibility?

If we accept the view that there is a harm–benefit *continuum*, and we have a moral reason to avoid causing unnecessary harm to others, we also have a moral reason to confer benefits to others, and this moral reason can become a positive obligation where the costs to ourselves are reasonable given the level of benefit. This is supported by the intuitive analogy between disability and enhancement. If disabling conditions constitute a *disadvantage* with respect to some relevant alternatives, enhanced conditions constitute an *advantage*. Moreover, one can easily imagine scenarios in which not enhancing a person's condition amounts to creating a disability. In an environment in which most people have had their long-term memory enhanced by 20 percent, people whose memory has not been enhanced are at a disadvantage in some contexts. Or, if a safe and effective vaccine against HIV/AIDS were to be developed, those not protected would be at a severe disadvantage.

But there are many objections to developing research programs aimed at enhancing conditions and capacities, and here we are going to review a few of them. In the bioethical literature, the press, and even in recent cinematography, enhancements are viewed with great suspicion.

Many worry about the safety of enhancement technologies and about the limited amount of knowledge even experts have about the consequences of, say, genetic engineering in certain domains. Although there might be very good reasons to decide against enhancement out of concerns about the safety of the procedures, this argument is not sufficient to show that enhancing is unethical. If science made significant progress, the safety of the procedures could be confirmed and the consequences of enhancement controlled. At that point, there would be no objection to proceeding with enhancement.

Another objection to pursuing enhancement stems from the tenacious view that the *natural* is good and the *unnatural* bad: that we should, in short, give priority to the natural over the artificial. The belief that the natural should have priority over the artificial, though common, has been found by many to be mistaken. In so far as naturally occurring foods are safer or healthier, there is a reason to prefer them; but in many cases, artificially prepared foods are safer and healthier. When natural processes are less costly or do less damage to the environment, there are reasons to prefer them. Thus, there is no reason to prefer a natural process to an artificial process in absence of other relevant considerations. These examples are supposed to show that the natural *per se* is morally neutral. Sometimes natural events are good, like a brilliant sunset or an abundant harvest. But often the natural does great harm (e.g. pestilence and floods) and can cause massive loss of human life.

One might characterize the practice of medicine (and science in general) as the comprehensive attempt to change the course of nature, because people *naturally* fall ill, are invaded by *natural* organisms like viruses and bacteria, and *naturally* die at a young age, often as babies. If we always prioritized the natural we would have to renounce the practice of medicine and the discoveries of medical science including vaccines and antibiotics.

 Exercise: Can you think of other debates where there is an uncritical acceptance of the view that what is natural is good?

There are two further objections that are often run together. One is the "playing God" objection, the idea that by enhancing in certain ways we are guilty of arrogance. The view goes like this: "Humans are not supposed to create better humans, because that would be arrogance on their part. They should just accept what God or Nature has given them without attempting to better it." This is not a very interesting objection, but it often leads to a second, more interesting objection. What are the consequences of enhancement? By intervening on the genes, we might change human nature and self-evolve. Is there something wrong with that?

One possible answer would rely on the view that the human species as it is should be preserved. This stems from the belief that there is something intrinsically good about being human. But is it really humanity as such that we value, the biological concept of it? Maybe what is valuable about humans is that typically they are also persons, with the capacity to be aware of themselves, make decisions for themselves in a rational and autonomous way and have complex feelings and emotions. The fact that all the persons we know are human is just an accident. If we did find those characteristics of persons in non-humans, we would (or should) still appreciate them and cherish them. These considerations might support the view that intrinsic value and moral status do not depend on the species individuals belong to, but to the fact that they are persons and, as such, have interests of a certain kind. Arguably, being human is neither necessary nor sufficient for having rights. If we recognize that what justifies granting rights to individuals is not the species they belong to but the interests they might have, then the question of whether to grant rights to post-humans is easily solved. The "human" in the phrase "human rights" is just supposed to highlight that differences of race, gender, and wealth are not relevant to the question whether someone should be granted those rights. If we take seriously the concern that some philosophers have for another kind of prejudice or discrimination, speciesism, then the "human" has to go and "human rights" will just be "rights of persons."

 Discuss: Do you think that rights should be reserved to humans as such?

Another set of objections stems from the societal implications of the widespread practice of enhancement. It is a common thought that some enhancing strategies such as genetic engineering are going to be very expensive and only the better-off in society will be able to afford them. As a consequence, the current divisions in society will become even less bridgeable. People who have better means will get further advantages over other people, such as better health, higher intelligence, additional talents, and so on. Although this is a serious concern, notice that it is not an ethical objection to enhancing as such, but a concern about fair distribution of resources. If we are concerned about whether enhancements will be fairly distributed, that means that enhancements are perceived as a good thing.

Some believe that the practice of enhancing and genetically engineering capacities will lead to a revision of our conception of agency. Agents typically enjoy a certain amount of freedom of action and are subject to judgments of praise for their achievements and of blame for their failures. But if the physical or intellectual achievement of the agent is only marginally due to effort and discipline and mainly produced by, say, a powerful drug, the achievement might no longer be a good reason to admire the agent. The argument is supposed to show that a pervasive use of enhancement might lead to a diminished sense of agency and responsibility. To assess the force of this argument one needs to be able to account for what the consequences of the practice of enhancement would really be for our conception of agency. Partly, it is an empirical question. We know what our current psychological reactions to illicit drug-taking by athletes are; we feel it is cheating. But the scenario in which everybody is given an opportunity to enhance their condition is significantly different and our reactions might reflect that change. It is not at all obvious that we would lose the sense of ownership of our own actions if the capacities that made it possible for us to achieve something desirable with our actions had been enhanced. One possible consequence of pervasive enhancement could be a "raising the bar" effect that would subtract little from the merits of the personal achievements of the individual.

That said, it seems as if the diminished agency objection is on to something. Suppose you are a runner and want to increase your speed by 20 percent. Also suppose that there are two methods by which you can achieve this target. You can take a pill that has an immediate enhancing effect on your speed or you can train for two months, three hours a day. (Notice that these are both *enhancing* strategies.) Now, you might have a morally relevant reason to prefer the hard way to the easy way. You might value self-discipline and think that you will grow as a person if you achieve this target by making a conscious effort to perfect your body during the next two months. You might believe that the sense of satisfaction you would get at the end of the training for having achieved the target is worth the time and the effort that are required. But all these valuable considerations do not amount to judging that it would be unethical for you to choose the easy option.

 Discuss: In what other circumstances can scientific progress affect social justice and alter the "human condition"?

This last objection invites further considerations about different types of enhancement. In the example of the runner who wants to increase her speed by 20 percent and has a choice, she can either train for two months, three hours a day, or take an enhancing pill. What is the more natural of these two strategies? "Natural" can refer to (1) the endowments we are born with and we have not acquired, (2) the features we regard as normal as opposed to those which result from illness, (3) the features which have remained the same as opposed to those that we have altered and (4) something that belongs to the world of nature and has not been processed or manufactured. One strategy is not obviously more natural than the other. It is true that regular training might not involve any of the artificial processing that might go into preparing the pill, but we do not seem to ethically object to this kind of artificial processing when we take medicines or eat food that we haven't grown ourselves. Of course we might have preferences for avoiding excessive processing and opt for so-called "natural" remedies or home-grown food when possible, but these preferences do not necessarily track moral reasons. We do not think that it is morally wrong to buy vegetables from a supermarket. And neither strategy is natural from the other points of view. Both strategies are aimed at the alteration of a feature (running speed) which is not innate but acquired. And running three hours a day is not something people normally do.

There can be enhancing strategies whose main aim is disease prevention and increasing life expectancy, and enhancing strategies that are aimed at improving cognitive abilities or physical appearance. While many feel that there is an intuitive difference between these different enhancement goals, it is difficult to draw a clear line between them from an ethical perspective. If it is not wrong to wish that our child were healthier, more intelligent or more beautiful it is difficult to see how it might become wrong to grant our own wish if we could.

The difference between the various aims or goals of enhancement lies in the risks that it would be worth running in pursuit of them. While for example considerable risks might be justifiable in order to cure a terrible disease or protect ourselves from almost certain death in a pandemic, it would be difficult to justify exposing our children to risks simply to change the color of their eyes or to make them better tennis players.

6.2.3.   Do we have a moral obligation to enhance?

Given the aspects of this debate that we have explored so far, there are at least three positions that could be defended on the basis of the acceptance of the harmed-condition view of disability and of the harm–benefit continuum. On the clear-moral-duty-to-confer-benefits view, no matter how slight the disability or insignificant the enhancement, parents have powerful reasons, which are always *moral* reasons, to minimize harm for, or confer benefits to, the person they are bringing into existence (subject of course to a safe method of achieving this and to the unambiguously beneficial nature of the proposed outcomes).

On the threshold view, parents have powerful moral reasons to prevent a disabling condition or enhance only when by not doing so they cause considerable harm to their children. In the context of reproductive choices, the parents' actions cease to be morally neutral and become subject to moral approval or condemnation when their actions have significant effects on the person they are bringing into existence in terms of harmed conditions that can be prevented or other conditions that can be enhanced. If the disadvantage caused by not preventing a disabling condition or not enhancing another condition is below a certain threshold, then there are no moral reasons to act. However, if the disadvantage is significant, then there are.

On the sliding-scale view, all actions are subject to moral scrutiny, not only those which have significant effects in terms of benefits and harms for future people. The idea is that parents always have powerful moral reasons to enhance and prevent disability, but there is an important difference between the clear-moral-duty option and the sliding scale. In the former, to confer benefits or to prevent harm is right and to withhold benefits or cause harm is wrong, no matter how slight the disability or valuable the enhancement. The reason to act is a moral reason and the degree of moral appraisal or condemnation for our actions co-vary with the degree of benefit conferred or harm prevented.

Therefore, the reasons to prevent the disabling conditions caused by *Holoprosencephaly* have a much greater moral impact than the reasons to enhance intelligence by say 15 percent. We are not talking here about the degree of strength of the motivating reasons, which can of course vary equally in the three options described, but about their being *moral* reasons in the first place.

 Discuss: Which of the described positions would you be prepared to defend? Why?

An advantage of the clear-moral-duty option is that it is a coherent and simple option that seems to be compatible with our conception of disability as a harmed condition and with the harm–benefit *continuum*. The problem with this option is that from a moral point of view some distinctions that many find intuitively strong (the distinction between preventing serious disability and enhancing for the purpose of minor benefits) are not transparent.

There are two serious problems with the threshold position. One is epistemological. It is never easy to measure how harmful or beneficial a condition is, as it cannot always be done out of context or inter-subjectively. This kind of calculation becomes even harder when applied to future people, whose interests and inclinations we ignore. The loss of a finger might be more significant for someone who may develop the ambition to become a great pianist than for someone who will have no interest in playing musical instruments, although it is disabling for both. The other problem is whether this option is really compatible with the harm–benefit *continuum*. If we recognize that a condition is harmful for someone, it would seem to follow that there are moral reasons to prevent that condition independently of the extent to which it is harmful. Can actions aimed at preventing harm ever be morally neutral?

The sliding scale option shares some of the epistemological problems that affect the threshold option but is overall more compatible with the harm–benefit *continuum*. What seems unattractive about this option is the conclusion that some reasons to act can be only *partially* moral reasons. Although it is perfectly reasonable to claim that we might have a variety of reasons to act in a certain way, it is more puzzling to believe that each of our reasons to act is only partially a moral reason.

Exercise: Apply the considerations that we have discussed above (e.g. societal repercussions of a new development or technology on fair distribution of resources, etc.) to a case of another ethically controversial objective of a scientific research program (e.g. life extending treatments).

6.3.   Ethical Constraints on Research Methods

When we think about the way the research is conducted and the consequences of experimental design and setting for the research participants, other issues come into play. There are moral reasons to make sure that the welfare and autonomy of persons are taken into account and that individuals are not harmed unnecessarily.

6.3.1.   Policy on deception in psychology

Is it ethical to deceive participants in social psychological research? Many authors have maintained that it is not (Kelman 1967; Bok 1999) and that the existing codes of ethics – which allow for the use of deception in some experimental circumstances – are in need of revision (Clarke 1999; Herrera 1999; Pittinger 2003). Should the use of deception be banned in social psychology research? Those who defend the legitimacy of its use, argue that the potential harms to research participants are not severe if the constraints set out by the current codes of practice are respected; and that the potential benefits that the experiments can produce for the research participants themselves and for society at large are of great ethical significance.

We shall review some of the arguments for and against the use of deception in psychological research. The codes of ethics of the American Psychological Association (APA) and of the British Psychological Society (BPS) allow for the use of deceptive or covert methods in psychological experiments, but they also set limits to the use of such methods and require that certain conditions be met when such methods are used. Although there are differences in recommendations made in the two codes, both codes require that deception be used only if:

1. There are no other effective procedures to obtain the desired experimental results.
2. The results are expected to be highly significant.
3. No physical harm or severe distress is caused to the research participants.

Recommendations are made that the experimenters debrief research participants as soon and as sensitively as possible after the experiment by giving them all the relevant information about the structure, the purpose and the value of the experiment. For all the details, we refer you to the *APA Ethical Principles of Psychologists and Code of Conduct* (2002), article 8.07, and the *BPS Ethical Principles for Conducting Research with Human Participants* (1992).

 Exercise: Are there experimental situations where conditions 1–3 cannot be met, but the use of methodological deception can be justified?

In spite of the many constraints placed by these professional codes on the use of deception in psychological research, some commentators have suggested that research participants are not sufficiently protected from the possible harms of deception and that the codes should be revised accordingly (Ortmann and Hertwig, 1997; Pittinger 2002). But what are the arguments against the constrained use of deception in psychology?

6.3.2. Objections to methodological deception

In many cases, the purpose of deceptive methods in psychology is to make sure that the research participants are not aware of what aspect of their behavior or their psychology is being studied. The methodological rationale behind this is that knowing that someone is studying how one behaves in certain circumstances can and often does affect the way one behaves in those circumstances. But the widespread use of deception in psychological research can be methodologically self-defeating. If deception were used in most of the experiments in a given discipline, and if the potential participants in the experiments were aware of this fact, any experiment in this discipline would generate suspicion in the participants. The participants would try to second-guess the experimenter and this would make the experimental results very difficult to interpret. Would the results indicate how people normally behave or would they indicate how people behave when they are trying to second-guess an experimenter?

Herbert Kelman (1967) warned against this possibility at a time when the use of deception was not as carefully regulated as it is today. But, thanks in part to constraints on the use of deception set out by the ethical codes of practice mentioned above, only some of the experiments performed by psychologists studying human behavior involve deception. For this reason, the risk of deception becoming a self-defeating strategy for methodological reasons is, at the moment, extremely low.

Probably the most influential argument against the use of deception in psychological research is that it violates the personal autonomy of the deceived participants and that the personal autonomy of research participants should never be violated. The argument can be developed by drawing an analogy between psychological research and other instances of scientific research involving humans, such as biomedical research. If psychological research involving humans must conform to the same standards that apply to biomedical research, for instance, then the principle of respect for personal autonomy demands that informed consent to experimental procedures is obtained by the research participants. In the studies involving deception research participants remain unaware of important details of the research in which they take part and therefore are not fully informed. They might be misled about the true objective of the research or about the role played by other actors in the experimental setting.

The argument we are considering comes in two versions, a stronger and a weaker one. The strong version says that deception always violates the personal autonomy of research participants and thereby it should always be banned in experimental situations, because full informed consent should always be sought. The weaker version, in contrast, draws attention to the fact that in biomedical research there are circumstances in which the potential subject is unable to give or deny consent concerning his or her participation in a given experiment. This can happen when the subject is an unconscious or critically ill patient, for example. In these cases, it is possible and legitimate to ask for consent on behalf of the patient from a legal representative or close relative after providing information about the details of the research protocol and the role that the patient will have in it. This is a form of *indirect* informed consent. Thus, in biomedical research, direct informed consent is not always necessary in order to respect the personal autonomy of the subject. The same, according to the view, could apply to psychological research. If obtaining direct informed consent were to compromise the value of the experimental results, indirect informed consent could be sought; if neither direct nor indirect informed consent were available, then the experiment would not be done.

Exercise: It is controversial whether informed consent is the best way to protect personal autonomy even in the context of biomedical research. What could be the arguments against it?

One thing to notice is the disanalogy between the medical research case and the psychological research case. The reason why direct informed consent is sometimes not a possible option in the biomedical case is that participants cannot give their consent (because of their state of health or lack of capacity). When we consider psychology instead, the reason lies in a methodological necessity, the fact that informing the research participant about the nature of the experiment would make the experiment useless. For the weaker version of the argument to go through, we have to concede that the difference between the two cases is irrelevant with respect to the requirements posed by the principle of respect for personal autonomy.

The measures already in place, one might argue, are already sufficient to guarantee the respect for the research participants' personal autonomy: research participants are informed before the experiment about the possible use of deception, they are debriefed with care and sensitivity after the experiment and they always retain the option to withdraw from the experiment at any time. The fact that the participants might not have all the relevant details about the purpose, design or setting of the experiment does not necessarily imply that the principle of respect for persons is dismissed or violated.

According to Alan Elms (1982), the use of deception in psychological research is likely to harm not only the deceived research participants, but also the researchers and their profession. The researchers may become morally corrupted as a result of using deceptive strategies and public knowledge about the existence of such strategies may undermine public trust in researchers in general. Why is the principle of confidentiality so important in medicine? Not just because it serves as a protection of the patients' right to privacy, but also because it helps establish and maintain good relationships between patients and medical professionals. In the same way, transparency might help the general public increase their trust in experimental psychologists and it might contribute to attracting more support for their research. On this view, a long-term consequence of the widespread use of deception in sensitive areas could be the public distrust of psychologists in particular and of scientists in general. If the relation of trust between potential experimental subjects and experimenters is systematically violated, experimenters might end up acquiring a bad reputation and the number of people wishing to participate in psychological research – and to fund through donations or taxation those institutions that promote such research – might diminish (Lawson 2001).

One might also object to methodological deception that experimenters using it risk becoming morally corrupt. But experimenters are aware that deception is a methodological tool. Their motivation for using experimental deception is simply a desire to conduct methodologically sound experiments. In contrast with other forms of human deception, there is no "wicked" motivation behind the use of experimental deception. Deception in this context is not motivated by the desire to defraud someone, or to get an unfair advantage over someone, etc. Its purely methodological nature makes experimental deception unlikely to have any effect on the personalities and moral dispositions of the experimenters.

 Exercise: Can you think of other contexts in which the use of deception is accepted?

The risk that experimental deception will give a bad name to psychology is also low, if experiments are conducted in a professional manner. As long as people in general and the participants in psychological experiments in particular understand that experimental deception is only an indispensable methodological tool and not the result of some "wicked" desire of the experimenter, no feeling of distrust towards the researchers is likely to emerge. This is especially the case if, as specified by current guidelines, after the experiment is complete, research participants are informed about the reasons why the experiment was conducted in the first place, about why some form of deception was methodologically necessary in this particular case, and about the potential value of the expected experimental results. In scrupulous debriefing, participants should be informed about the way risks were avoided, should be allowed to ask questions and reassured of the value of the research. This will often generate a feeling of satisfaction from having contributed to obtaining important results. Moreover, as we shall see, even when what the participants learn about themselves can be distressing, these potentially harmful effects can be mitigated by the awareness that the form of behavior exhibited by them during the experiment is by no means an isolated occurrence.

6.3.3. Benefits of methodological deception

The whole debate about whether deception is morally permissible resulted from a reflection on the potentially harmful consequences of the use of deception on research participants. The classical example cited in the context of a discussion of potential long-term harmful psychological effects is the notorious experiment conducted by Milgram on people's tendency to obey authority. The participants to this experiment were recruited by being told that they were going to be part of a study on memory and on how punishment affects learning. In the lab, each participant was told that he or she had to play the role of "teacher" and that another participant present in the room had to play the role of "learner." Unknown to the genuine participant, the learner was in fact a confederate of the experimenter. The teacher was supposed to ask questions to the learner and, in case of incorrect answers, to administer electric shocks of progressively higher voltage by manipulating a simple electronic device. Unknown to the participant, the learner did not in fact receive any electric shock. The learner only pretended to be in pain when the participant "administered" the electric shock. And the learner's manifestations of pain were proportional to the voltage of the electric shock that the participant believed to be administering. In general, when the learner's complaints about the pain became relatively strong, research participants manifested their uneasiness with (what they believed) was happening to the learner. Many of them asked the researcher to stop the experiment. In response to these requests, the experimenter demanded obedience, insisting that it was very important for the teacher to follow the instructions independently of how loud the learner's screams were. In the end, 65 percent of the participants inflicted (what they believed to be) electric shocks of the highest voltage to their respective learner, in spite of the learner's pleas to stop (Milgram 1974).

 Discuss: Do you think that it was unethical to conduct the Milgram experiment?

Many commentators today consider the Milgram experiment a paradigmatic example of the unethical use of deception in psychological research. The participants were deceived about the purpose of the experiment and about the role of the experimenters and of the learners.

Moreover, they were put in a very distressing situation due to the experimenters ordering them to follow the instructions despite their uneasiness. After debriefing, the participants had to deal with the knowledge of the fact that they had been capable, under the influence of authority, of inflicting considerable pain to innocent human beings. The standards for debriefing at the time were not the same as they are today and it is possible that many participants did not receive an adequate explanation of the nature of the experiment and of its rationale. It is possible that the debriefing offered at the time did not help all the participants to deal with the psychological discomfort generated by the realization of what they had done.

The Milgram experiment is often taken to have caused significant psychological harm to the research participants. However, the significance of the harm actually caused to the participants remains controversial. Elms (1982), who worked behind the scenes of the experiment and interviewed the participants after their experience, claims that they suffered remarkably little harm given what he had expected after witnessing their reaction during the experiment. According to him, the experience had been distressing for them, but no more than an emotionally involving movie or a disappointing job interview.

Whether the experiment produced significant psychological harm on the participants is a difficult empirical question, one that cannot be answered by appealing to untested intuitions or casual observations. In fact, it is one of those questions that experimental psychology is designed to answer. Very likely, it is one of those questions that could be answered only by using deceptive methods in experimental settings. We do not actually know whether Milgram-like experiments produce significant and long-lasting psychological harm on research participants. If they do, then there are moral reasons for not conducting such experiments. But if the experiments do not produce any significant psychological harm, there are no such moral reasons. Moreover, the presence of moral reasons against conducting an experiment is compatible with the presence of moral reasons in favor of conducting the experiment. Thus, the fact that some experiments that use deception generate some level of psychological stress in their participants does not by itself entail that the experiments should be banned.

What moral reasons could there be in favor of doing these experiments? Deception regarding the main purpose of the experiment is used to avoid the so-called Hawthorne effect, i.e. the tendency of the research participant to behave in accordance to what they think the experimenter's expectations are (Gillespie 1991). In social psychology, though, where often the object of the research is a form of undesirable behavior, the opposite effect can occur. For example, if research participants are made aware that the object of the experiment is aggressive behavior, then they might become self-conscious and actively try to refrain from engaging in aggressive behavior for the duration of the experiment.

Psychological evidence suggests that reliable data about how people behave in certain situations cannot be obtained by simply asking them how they did behave or how they would behave in similar situations. People are often mistaken about their behavioral tendencies and the ways in which they describe themselves or revise their self-descriptions on the basis of evidence are usually biased by their (unconscious) desire to fit this or that particular profile. It follows that there are circumstances in which it is virtually impossible to study a phenomenon without using some form of deception. Suppose one is studying helping behavior in emergencies. What people say about what they would do in emergency situations is a very poor guide to what people actually do in such situations. But real emergencies do not occur on request, nor is it permissible to intentionally provoke them. Therefore, helping behavior in emergency situations can only be studied by means of simulated emergencies and simulated emergencies are a form of deception. Consider the Good Samaritan experiment (Darley and Batson, 1973), in which the researchers wanted to demonstrate that altruistic behavior is affected by external and contextual factors rather than by personality traits. They aimed at showing that people in a hurry are much less likely to offer assistance in emergencies, independently of their personality. Darley and Batson devised an experiment in which some seminarists were told that they had to reach a building in a hurry to complete a task (ironically, talk about the story of the Good Samaritan). Participants found a man slumped in an alleyway on the route to the building, but most of them did not stop to help him. Those who believed themselves to be late exhibited less altruistic behavior than those who were not as rushed.

In other words, there are many experimental situations where deception is methodologically necessary in order to obtain reliable results. This does not by itself mean that deception in this context is morally permissible. As geneticists know very well, controlled (as opposed to natural) breeding experiments would be methodologically necessary in order to obtain reliable data about the effects of human genes on human phenotypes. But, in spite of this, such experiments are not morally permissible. In order to determine whether experimental deception is morally permissible one needs to determine what are its moral benefits and its moral costs and then judge whether the benefits outweigh the costs.

6.3.4. Deception in psychology and beyond

One argument present in the literature says that, given that the use of deception is widespread in everyday life and in other fields of investigation, then it should not be banned in social psychology. It has been argued that since the use of deception is so generally accepted in marketing research and in the labor and housing markets, it should also be accepted in social psychology research, because there are no good reasons to apply different standards to the two cases (Kimmel 2001; and Riach and Rich 2004).

However, it is not sufficient to claim that deception is used in a number of other contexts to show that its use in social psychology is justified. First of all, more subtle comparisons should be made in order to determine whether the use of deception in social psychology research is significantly similar to the use of deception in marketing research and in the housing and labor markets with respect to the relevant variables (magnitude of costs and benefits, etc). Moreover, the fact that one practice is widely endorsed or tolerated does not make it ethically justified. This notwithstanding, there are cases in which a fruitful analogy can be drawn between the use of deception in social psychology and the use of deception in other fields.

Ethical considerations in favor of deception come from a utilitarian perspective. Many valuable experiments that involve deception do not cause any discomfort or harm to the research participants. For those experiments that do cause some degree of discomfort or harm, there are many cases where the significance of the results outweighs the significance of the potential discomfort or harm caused to the research participants. In social psychology, methodologically sound research that requires some form of deception can have great utility both for the research participants involved and for society at large. Good research can contribute to identifying some patterns of behavior that negatively affect both the person who exhibits them and others around them who experience that behavior. Awareness of potentially disruptive behavioral dispositions can help develop strategies to avoid their unpleasant effects.

The results of social psychology research can not only help social scientists to gain a better understanding of human behavior and of the way in which human societies work. They also increase the self-awareness of each person who participated in the actual research. Let us recall the famous studies by Stanley Milgram on obedience to authority (which we mentioned briefly in chapter 1). It is not difficult to see the potential value of the experimental results that Milgram obtained, especially if one considers the historical context in which he became interested in obedience. After the end of the Second World War, many questions were asked about the psychological and social mechanisms that led a great many people to be involved in activities of ethnic cleansing. Milgram wanted to test his hypothesis that people have a tendency to follow authority even when the actions they are asked to perform go against some of their core personal values. His results brought general awareness, in the scientific community and in some sections of the general public, that the influence of authority can lead one to act in ways that one believes to be wrong or immoral. This is important because such awareness may help individuals and societies prevent events such as the Holocaust from occurring in the future.

As we previously said, if an experiment is likely to cause some discomfort or harm to the participants, there are powerful moral reasons against conducting the experiment. But, when the harm and discomfort for the participants are not severe, the moral reasons against conducting the experiment can be outweighed by the moral reasons in favor of conducting it. It can be argued that if the psychological harm to the research participant surpasses a given threshold, then the experiment is not morally justified. This conclusion can be drawn independently of the potential benefits for the research participant or humanity as a whole, because doing the experiment would constitute an act of injustice against the research participant. This argument constrains the utilitarian calculus of the costs and benefits of certain experiments by saying that the costs for the research participant cannot be higher than a certain amount. But this argument does not entail that there should be a generalized ban on the use of deception in experiments. Instead, rather than a ban on deception, such argument suggests that a sensible threshold of psychological harm to the research participant should be identified and that regulations ensuring that such threshold is never surpassed should be designed. It is noteworthy that the existing ethical codes insist that in no case should an experiment be done if that experiment is likely to cause severe psychological harm to the participant. That is, the existing regulations attempt to ensure that a sensible threshold of harm is never surpassed and, thereby, they seem to be in line with the proposal we have just discussed.

The study of prejudices and biases against individuals of a particular race, gender, age, sexual preference, or physical appearance can help uncover aspects of human behavior which cause unfair discrimination, which individuals and society can then try to control or change. That is why results in this area of investigation have great significance. Data about the relevant behavioral tendencies can be obtained in laboratory situations as well as in the field.

Many questions emerge about the implications of the use of deception in the psychology laboratory. Is there some important interest of the research participant that researchers set back by first using deception and then reporting the results of their study? Acquiring knowledge of one's own negative psychological traits can be distressing. On the assumption that individuals have an interest in avoiding psychologically distressing situations, it follows that research participants have an interest in not acquiring knowledge about their own negative psychological traits and, thereby, they have an interest in not being deceived by researchers aiming at identifying and studying negative psychological traits. But the existence of this interest by itself does not show that the use of deceptive methods in psychological experiments is illegitimate or ethically problematic. Not all interests are legitimate and not all interests have the same ethical significance. A bias against overweight job applicants generates an unfair advantage in favor of normal-weight applicants. Moreover, the bias harms companies, since it leads employers to prefer to hire less competent normal-weight individuals over more competent overweight individuals. Employers who are aware of the existence of this bias may be able to avoid the unfair dismissal of overweight candidates. In this way, they may be able to enhance the efficiency of the hiring process and the prospects of success of their company as well as to positively contribute to social justice. Knowledge about the existence of the bias could also prove useful in the creation of anti-discriminatory legislation and of other corrective mechanisms. Thus, the interest that research participants might have in remaining ignorant about their unconscious discriminatory biases against overweight people is less ethically significant than the interests that research participants and society at large have in knowing about the existence of such biases.

The ethical significance of the research participants' interests in remaining ignorant can also be challenged from a different direction. It is possible that, in some individuals and in some circumstances, learning about one's biases, limitations, or other negative psychological traits might result in lower self-esteem and might have negative consequences on one's future happiness. But, in general, if the experimental results are properly explained and understood, this is unlikely to happen.

First of all, one needs to consider that in the cases we have discussed, and in many other similar cases, the negative psychological traits uncovered by investigators are present in a very large section of the population. Such traits are, in other words, statistically normal. Moreover, learning that someone is prone to make certain kinds of reasoning mistakes or to engage in certain kinds of discriminatory behavior can have a positive effect on one's self-conception. Such knowledge gives people the possibility to try to overcome their limitations. And the fact that one is in a better position to overcome one's limitations may improve rather than damage one's self-esteem.

Some authors have argued that covert and deceptive methods might be acceptable in watchdog reporting and in police, military, and espionage activities but not in social science research because the social scientist has more responsibility to the rest of the community (Erikson 1967, p. 367). Although there are relevant differences between social psychology research and, say, watchdog journalism which might affect the ethical justification of the methods used in these activities, the use of deception in social psychology experiments might be actually seen as *less* ethically problematic than the use of the deception in at least some of these other contexts.

Let us focus on the case of watchdog reporting. One reason why deception is by some people perceived as more problematic when it occurs in social psychology research than when it occurs in watchdog journalism may have to do with the fact that (at least in people's perceptions) the potential benefits of experimental psychology have less immediate impact than those of watchdog journalism. But, even if the results of psychological research often do not have as much front page potential as the findings of watchdog reporting, such results can be as beneficial to society as those of the best watchdog reporting, if not more. Thus, those who agree that deception is a legitimate tool in professionally done watchdog reporting should also agree that deception is a legitimate tool in well-conducted psychological experiments.

Firstly, the conduct of social psychology researchers is more strictly controlled than the conduct of undercover reporters. The reason for this is that, on the current regulations, researchers are required to obtain permission to proceed with their experiments from ethics committees and they are required to obtain permission to use the data generated by each participant from the participants themselves at the moment of debriefing. In contrast, at this stage, no ethics committees exist with the purpose of examining watchdog investigations before they take place and to stop them when necessary – although the reporters are liable to criminal prosecution if they trespass private property and engage in other similar activities. One must also notice that, even when researchers cause some psychological distress to the research participants, the researchers are actually benefiting the research participants by allowing them to discover some of their own psychological traits and giving them the opportunity to acquire some form of control over such traits. The same is not true of watchdog reports, whose purpose is usually to show that someone is knowingly doing something that should not be done.

 Exercise: Can you think of other concrete cases in which the principles of respect for personal autonomy and beneficence could conflict?

 Discuss: Do you think that scientific research should be regulated more severely than other human activities?

6.4.   Ethical Constraints on Scientific Research

In the previous sections we saw how ethical issues might arise both from the goals and methods of scientific research. Should research be regulated in a way that marks it as different from other human activities?

Looking at the various reasons which militate in favor of ethically regulating research it becomes clear that ethical issues can be raised by almost any research activity, but the extent to which they affect the interests and rights of the individuals involved can vary considerably. But as there are perfectly respectable instances of scientific research that do not seem to be candidates for ethical regulation, there also seem to be activities that do not count as research and are morally problematic (such as extended surveys for the purposes of policy which involve sensitive personal data).

Apparently, the class of activities raising ethical issues and needing ethical regulation does not coincide with those activities that are considered as research. Presumably, the reason for this is that what really matters from an ethical point of view is that the rights of the individuals involved in the activity are safeguarded and their interests taken into careful consideration. When we are thinking about which activities should come under ethical scrutiny, the main issue should not be whether they count as scientific research according to some demarcation proposal, but whether the interests and rights of individuals are likely to be affected. Another way of approaching this issue is to say that from an ethical point of view, some activities should be reviewed in the same way as research is, even if they are not aimed at extending knowledge via a scientifically acceptable method. This does not mean that they are research, but that they should be monitored for the effects that they might have.

Let me offer an illustration of this point. The use of innovative, non-validated drugs in a therapeutic setting might not count as research, if contributing systematically to a body of knowledge is not its main purpose. However, in some cases ethical reviewing might be appropriate. If the risks are high, one could argue that an independent review board (IRB) should assess the use of non-validated drugs carefully even if it does not count as research. Therapeutic freedom of choice is one of the main elements of the medical profession. However, if physicians apply a non-standard procedure that can be dangerous for the patient, their activity might be as ethically problematic as an instance of research. Physicians might be unrealistic in their evaluation and, as a result, cause unnecessary harm.

Similarly, if we think that some non-human animals deserve direct moral consideration and should not be confined or made to feel pain unnecessarily, research involving non-human animals should be regulated in such a way as to reduce the potential for frustrating these basic interests. But other human activities involving animals (e.g. intensive farming) should also be regulated on the same basis. The mere fact that other human activities do not count as instances of research should not excuse them from ethical regulation.

Summary

In this last chapter we addressed one aspect of the complex relation between science and the rest of society: whether the aims of current scientific research and the procedures by which these aims are pursued should be constrained by ethical regulation. We explored these issues by reference to the details of two very lively debates: one on the ethics of enhancement in the context of assisted reproduction, and the other on the legitimacy of methodological deception in psychology.

In the course of our examination of science, we challenged the presumed methodological distinctions between natural and social sciences, and between the rationality exhibited by scientists and that exhibited in everyday thinking and theorizing and in other areas in which knowledge is acquired, systematized and revised. If the argument for continuity is successful, then it seems that there are no grounds to believe that scientific research should be more severely constrained, from an ethical point of view, than any other human activity that is likely to impact on the rights and interests of those individuals involved. In so far as science is a powerful institution in our societies, its work needs to be monitored and reviewed as the work of any other powerful institution. But there seems to be nothing intrinsic to scientific practice that makes it more dangerous than any other human practice.

On the contrary, science seems to be invested of a special responsibility because it is often by doing science that we can further our aims: curing diseases that cause suffering; establishing which individuals can feel pain in order to protect their interests; uncovering our implicit tendencies to racial and sexual discrimination; identifying areas in which our rushed judgment causes injustice; preventing disabling conditions; and the list can continue. In this respect, science is to be promoted, because it is a way for individuals and societies to achieve moral progress.

Issues to think about

1. In what sense do scientists have an ethical responsibility towards the rest of society?
2. Is there any personal right that cannot be violated in scientific research?
3. Should research in countries with radically different cultures from ours be conducted in accordance with our moral principles?
4. How should we resolve the conflict between the respect for autonomy and beneficence?
5. How can society be made to trust science?

Further resources
Other ethical debates which take issue with the objectives (e.g. human reproductive cloning) or methods (e.g. use of animals in biomedical research) of some research projects can also be characterized in terms of there being ethical or other societal constraints on the practice of science. Abundant literature can be found on ambitious research goals and the ethical debates that accompany those attempts. Some general web resources are the ESRC Science in Society site (www.scisoc.net/SciSoc/), and the selection of pod-casts in the Science and Society site (www.scienceandsociety.net/podcasts/). In the thematic bibliography you find a list of readings on some specific issues that have generated lively bioethical debates, but additional resources can be found by accessing reports aimed at policy makers (often available on the web) and by consulting reference texts and collections of papers on specific areas (genethics, animal ethics, computer ethics, ethics of research in developing countries, etc.).

Whether science has a contribution to make to the rest of society by creating opportunities for doing good, is also a very interesting question to ask, and neglected in the literature. As a taster, I recommend the controversial paper by Harris (2005) on whether there is a moral obligation to support or even participate in scientific research.

Conclusion: Science as an Activity

The picture that emerges from the careful consideration of methodological, epistemological, ontological, and ethical issues concerning the practice of science suggests continuity between scientific research and other human activities. Here we have only had a glimpse at many of the debates that have put into question the special status of science. Can we really point at what makes a human activity an instance of scientific research? When scientists defend a particular hypothesis, do they employ argumentative strategies that are not available to philosophers or laypeople? Do scientific theories capture the essence of reality, or do they just offer one explanation of phenomena that is suitable for some specific purpose? Is change in science always based on rational objective principles? Does it always lead to progress? Should scientific research be regulated more severely or more liberally than other human activities which affect morally relevant interests?

In ordinary language, and in the philosophy and sociology of science, "science" (or "scientific") may refer to different things:

1. The *end-product* of the application of a method of investigation to a certain domain. In this sense, "science" refers to a body of knowledge with certain characteristics.
2. The *process* of acquiring knowledge in a way that is sensitive to evidence and open to rational criticism. In this sense, "science" refers to a method of investigation.
3. An entire *community* engaging in the activity defined by (1) and (2), including the people who do research, the institutions where they work, the laboratories where they conduct experiments, the journals where they publish their work, the books they write, etc. In this sense, "science" refers to an institution, an integral part of many human societies.

In this introduction we have examined some aspects of each of these meanings of the word "science." Although considerations relevant to (1), (2), and (3) cannot be easily kept apart, there has been a progression from questions about method, to the consideration of science as a body of knowledge and finally to the pressures on and the values of science as an institution. In chapters 1, 2, and 3, we have concentrated on the method of science, the process by which scientific knowledge is acquired, and the way in which theories are structured, confirmed on the basis of evidence and used in explaining interesting phenomena. In chapters 4 and 5, we have turned to science as an ever-developing body of knowledge whose language changes together with the acceptance and rejection of theories, and whose progress is not always easy to measure. The function of scientific theories has also been examined: are they supposed to give us an insight in how things *really* are, or are they useful tools for prediction? Finally, in chapter 6, we have considered science as a community of researchers whose work is often constrained by, or has a valuable contribution to make to, issues that affect the morally relevant interests of individuals in society.

Have we succeeded in isolating a set of special features that science alone displays? We haven't been successful in this task, but for very good reasons. What we have found is significant continuity between scientific research and other human activities, between natural and social sciences, and even between science and philosophy. But to stress these elements of continuity and the difficulties in demarcating science is not to render the labels "science" or "scientific" obsolete or useless. On the contrary, only after reflecting on the elements of continuity between science and non-science, can we appreciate the many dimensions of scientific research and start grasping the often implicit values that we refer to when we talk about something being a science, or being scientific. Although "science" and "scientific" do not appear to be anything more than descriptive terms, we often use them as evaluative terms. When we call a certain activity an instance of scientific research, or when we refer to a discipline as scientific, we might want to highlight that they meet certain standards of rationality or systematicity, or we might imply that their scope is limited with respect to their aims or their capacity to take into account relevant human interests. "Evolutionary biology is science" can be a judgment of the intellectual respectability of the activities involved in the development of evolutionary theory, or a defense against those who criticize the subject as an unsupported hypothesis. But to say: "That's only science" often indicates that matters of fact that we can investigate empirically via the methods of science do not exhaust the scope of our interests.

We do not just describe what an activity or a discipline are like when we employ these labels, but we also assess them on the basis of values that depend upon the interests that we, as individuals or as societies, have in pursuing a certain investigation of the natural and social context in which we live and operate. In the last chapter we discussed in some detail the ethical constraints that apply to specific instances of scientific research, and how objections can be raised to some research objectives or research methodologies on the basis of the impact they might have on morally relevant interests that we want to safeguard.

But it can be argued that, whereas some human interests can conflict with the interests of some specific scientific projects, other human interests cannot be adequately safeguarded without promoting science. By letting science flourish, we guarantee the constant contribution of the scientific community at large to a deeper understanding of the problems that we need to face and to what might enhance the well-being of individuals and societies.

The practical contribution of science to the rest of society is obvious in some fields: biomedical research makes a very visible impact on the development of treatments for debilitating diseases. But there are other benefits that apply more generally. The availability of empirically grounded scientific hypotheses allows us to support empirical claims in arguments about issues that affect the whole of society. The availability of those hypotheses, as fallible as they might be, means that we can justify empirical claims on grounds that are open to rational assessment and criticism. Are psychopaths morally responsible for their actions? Are free-range chickens happier than battery chickens? Are armed conflicts ever successful in bringing about democracy? These are value-laden questions about some of the issues we care about, as individuals and as a society, but they cannot be answered without an analysis of some of the concepts involved (e.g. happiness, moral responsibility, democracy) and, crucially, a careful assessment of empirical claims that are either supported or rejected on the basis of scientific investigation (e.g. whether confinement as opposed to roaming freely affects negatively the physical and psychological well-being of chickens).

The progress that science allows us to make is not measured exclusively in terms of its giving rise to technological advances or a series of breakthroughs in medicine. It is also measured in terms of the justification that science makes available for the beliefs that are the basis for our social interaction and our understanding of the surrounding environment.

Glossary

The following glossary contains basic definitions of relevant philosophical terms and a brief introduction to some of the views, philosophical movements, and authors mentioned in the text.

a posteriori   The truth of a statement is known *a posteriori* if it is known on the basis of experience.

a priori   The truth of a statement is known *a priori* if it is known independently of experience.

ad hoc   Latin expression which means: "for this purpose." An explanation is "ad hoc" if it involves hypotheses that are introduced with the sole purpose of saving a theory from falsification.

analytic   A statement whose truth or falsehood depends on its logical structure or the meaning of the terms it contains. E.g. "All tall buildings are buildings"; "Cats are animals."

anomaly   Literally, "deviation from the norm." Used in philosophy of science to indicate the incongruity between an observed event and what the current theory had predicted.

anti-naturalism   In the context of the debate about the status of social sciences, the anti-naturalist argues for a significant discontinuity between the methodologies and aims of natural and social sciences and claims that social facts must be explained in terms of meaning, purpose, or interpretation, and not in terms of causal relations and laws of nature.

Aristotle (384–322 BC)   Greek philosopher who vastly contributed with his teaching and extensive writings to the foundations of logic, metaphysics, epistemology, ethics, political philosophy, and also biology, physics, and astronomy.

autonomy (principle of respect for personal autonomy)   Autonomy is self-governance and self-determination, the capacity agents have to form beliefs and intentions, make decisions and act on the basis of reasons that reflect their values, without being coerced. The principle of respect for personal autonomy says that we have an obligation to respect the beliefs, choices, and actions of autonomous agents. What the principle involves is controversial, but it is often seen as the basis for practices aimed at protecting confidentiality and eliciting informed consent.

axiomatic system   An axiomatic system contains: undefined primitive terms; defined terms; axioms (statements accepted without proof) and theorems (statements subject to proof).

axiomatization   Axiomatization is the attempt to capture the structure and content of a scientific theory into a formal system of statements. Some of these statements contain undefined primitive terms (axioms) and the other statements (theorems) are deductively derived from them. Axiomatization requires characterizing with some precision the domain of the objects posited by the theory, providing a list of primitive terms, providing rules of composition of well-formed formulas, establishing which statements are axioms, etc.

Ayer, Alfred (1910–1989)   British philosopher who divulged the fundamental theses of the logical positivist movement in his very influential *Language, Truth, and Logic* (1936).

Bacon, Francis (1561–1626)   Philosopher and politician who was responsible for challenging Aristotle's ideas and codifying inductivism as the basis for scientific methodology. His seminal methodological work, *Novum Organum*, appeared in 1620.

Bayes's Theorem   Bayes's theorem has been used to formalize subjectivist accounts of evidence and aspects of confirmation of scientific theories. It says that the probability of a hypothesis *H* conditional on a given body of data *E* is the ratio of the unconditional probability of the conjunction of the hypothesis with the data to the unconditional probability of the data alone.

Bayesianism   The view that the notions of scientific justification and confirmation can be captured by Bayes's Theorem.

beneficence (principle of)   Beneficence literally means "doing good." The principle of beneficence says that we have an obligation to do good, that is, to benefit others and to prevent and remove harm from others. Both the exact formulation of the principle and its implications are a matter of controversy in ethics.

Black, Max (1909–1988)   Philosopher of language, mathematics, and science influenced by Frege, Russell, and Wittgenstein. He argued that induction can be inductively justified at no risk of circularity.

Bohr, Niels (1885–1962)   Physicist responsible for developing a model of the structure of the atom on the basis of elements of Planck's quantum theory. Author of *Atomic Theory and the Description of Nature* (1934).

Boyd, Richard   Contemporary philosopher of science and of mind, mainly known for defending scientific realism. Author of many articles and an anthology with core readings: *The Philosophy of Science* (1991).

Boyle, Robert (1627–1691)   Scientist who contributed to both pneumatics and chemistry and supported in his writings a mechanistic explanation of nature and an experimental approach to science. He was one of the founders of the Royal Society.

brains in vats   Skeptical scenario made famous by Hilary Putnam, who imagines an evil scientist removing people's brains during the night and placing them in vats. The scientist connects the brains to a sophisticated sensory stimulation machine that reproduces the sensations these brains would receive if still embodied and interacting with the external world.

Carnap, Rudolf (1891–1970)   Leading member of the Logical Positivism movement, he contributed significantly to a variety of issues in the philosophy of science, endorsed the classic view of theories and argued for the possibility of providing a translation of theoretical statements into observation statements via correspondence rules. He is the author of *The Logical Structure of the World* (1928), *The Logical Syntax of Language* (1934), and *Philosophical Foundations of Physics* (1966).

causal theory of reference   According to this theory of reference, a name/natural-kind term acquires its referent via an initial baptism, and later refers to that entity by bearing a causal connection to the original baptism, independently of the concepts speakers associate with it.

circularity   An argument is circular if one of the premises is identical to the conclusion or if the premises are such that we are in no position to know them unless we already know the conclusion.

confirmation   Any process by means of which the probability that a scientific hypothesis or theory is true increases. Usually consists of observations that are compatible with the predictions that can be made on the basis of the hypothesis or theory in conjunction with auxiliary hypotheses.

constructive empiricism   View developed by Van Fraassen which differs from scientific realism as it denies that current theories are true or approximately true, but also differs from instrumentalism as it denies that theories lack truth-aptness. The constructive empiricist believes that there is no good independent justification for believing that theories are true and that we should just accept current theories as empirically adequate.

contingent   Opposite of "necessary." Something that is, but could have not been. It is contingent that I wrote this book (I could have not written it).

conversion   In Kuhn's account of scientific change, the shift from one paradigm to another requires conversion, because it cannot be motivated by rational argumentation based on evidence. This is because there is no evidence that can be neutral with respect to competing paradigms and there are no reasons to prefer one paradigm over another which are paradigm-independent.

Copernicus, Nicolaus (1473–1543)   Scientist who is responsible for defending systematically the heliocentric theory of planetary motion and supporting it with detailed astronomical evidence in his treaty *Of the Motions of the Heavenly Bodies* which appeared in 1543 and is thought to have kick-started the Scientific Revolution.

correspondence rules   Statements containing both theoretical and observational terms which are supposed to provide a formal system with empirical content and therefore allow the theorems of a theory to be first interpreted and then tested. Whether correspondence rules are analytic (as definitions) or synthetic is an object of controversy. Here is an example (Carnap 1966): "If there is an electromagnetic oscillation of a specific frequency, there is a visible greenish-blue color of a certain hue." A correspondence is established between a theoretical term ("electromagnetic oscillation") and an observational term ("greenish-blue color").

corroboration   This notion, introduced and developed by Popper, refers to the past performance of a theory, and in particular the extent to which it has survived critical discussion and severe tests.

Darwin, Charles (1809–1882)   Naturalist who initiated a scientific revolution with his views, defended in *The Origin of Species* (1859), about the evolution of species. In *The Descent of Man* (1871) he argued that humans and other primates have a common ancestor, generating strong reactions in the popular press of the time.

deduction   Mode of inference in which the truth of the premises is intended to guarantee the truth of the conclusion.

deductive nomological model   View of scientific explanation put forward by Hempel to account for the logical relation between an event to be explained and the factors that contribute to its explanation, when the event is the conclusion of a deductive argument which has the lawlike statement and the initial conditions as its premises.

deductive statistical model   View of scientific explanation put forward by Hempel to account for the logical relation between a statistical regularity to be explained and the statistical laws and initial conditions that contribute to its explanation, when the regularity is the conclusion of a deductive argument which has the statistical law and the initial conditions as its premises.

deductivism   Style of reasoning contrasted with inductivism and defended by Popper as the method of science. According to this view, we move from the general to the specific, and evaluate general statements by deriving from them hypotheses that can be subject to testing.

demarcation criterion   Systematic account of what makes science different from non-science or pseudoscience. The logical positivists thought that verifiability could provide a demarcation criterion, whereas Popper argued that falsifiability had a better chance. After the social turn in the philosophy of science, many have suggested that factors external to scientific method are relevant to what is regarded as a science and what isn't.

deontology   Approach to ethics according to which whether an action is right or wrong does not depend on the consequences that ensue from that action, but on whether it respects certain general principles (duty-based ethical views were defended in a systematic way by Immanuel Kant). Rights-theory is often embedded in a deontological approach, and tells us that we have special obligations towards certain individuals on the basis of their status.

Descartes, René (1596–1650) Mathematician, physicist, and philosopher, who broke with the Aristotelian tradition and attempted to develop new theories of space and motion. He was interested in scientific and philosophical methodology as well, and greatly influenced thinkers of his time. Author of *Meditations* (1641), *Principles of Philosophy* (1644), and *Passions of the Soul* (1649).

descriptivism In philosophy of language, this is the view that terms get their reference fixed by the descriptions that speakers associate with those terms. Frege and Russell are taken to be descriptivists.

Duhem, Pierre (1861–1916) Physicist, mathematician, and philosopher of science who wrote about the relationship between observations and theory and argued that failed predictions could falsify the theory tested as well as the auxiliary hypotheses necessary for the testing (Duhem–Quine thesis). He also had an instrumental view of scientific laws. Author of *The Aim and Structure of Physical Theory* (1906/1914).

Einstein, Albert (1879–1955) Physicist with an interest in philosophy and cosmology. He developed the special theory of relativity guided by the intention to reconcile Newtonian mechanics with the laws of electromagnetism.

empathy Different notions of empathy have been formulated, but a general definition can point to the way in which one can understand the beliefs and emotions of another individual not through the mediation of a theory, but by putting oneself in that individual's shoes. Empathy is thought by some to have an important role to play in explanation in the social sciences.

empirical adequacy A theory is empirically adequate if it is compatible with the available evidence and has not been refuted by it.

empiricism (adj. empiricist) Opposite to rationalism. View according to which all knowledge comes from sense experience.

enhancement Any strategy or treatment aimed at improving existing conditions. Often used to refer to genetic enhancements which can apply to a variety of conditions and capacities, and can be exemplified by immunity from certain illnesses or improved cognitive abilities.

entailment Logical relation which is converse to that of consequence. My having a sibling is a consequence of my having a sister. That I have a sister entails that I have a sibling.

entrenchment Feature of a predicate which is embedded in everyday discourse and is used to describe objects and make predictions about future observations.

equant and epicycles Mathematical concepts developed by Ptolemy in order to account for the apparently anomalous motion of some celestial bodies in his geocentric model of the universe.

essentialism The view that some things (e.g. people, natural kinds etc.) have core properties that make them what they are and are often hidden or not superficially assessable.

expressive   A statement is expressive if it is the manifestation of the desire or preference of an individual or a group. Expressive statements are not true or false in virtue of how things objectively are and (according to some logical positivists) have no cognitive significance.

falsifiability   Popper's criterion for regarding a theory as scientific. A theory is falsifiable when it is possible for the theory to generate predictions that are disconfirmed by evidence in rigorous testing.

falsification   A hypothesis is regarded as false when at least one prediction made on its basis has been disconfirmed by evidence.

Feyerabend, Paul (1924–1994)   Philosopher of science with a provocative agenda and an engaging writing style who argued against the supremacy of science over other traditions of thought, endorsed incommensurability, and explored the consequences of the privileged role of science in democratic societies. Author of *Against Method* (1975).

Frege, Gottlob (1848–1925)   Logician and mathematician who is considered one of the founders of analytic philosophy. He is responsible for breaking with Aristotelian logical tradition and developing a new quantificational logic. He also contributed vastly to the philosophy of language by discussing the relation between sense and reference. He endorsed descriptivism in "On Sense and Meaning" (which appeared in German in 1892).

Fresnel, Augustin (1788–1827)   One of the founders of the wave theory of light. Mathematician and civil engineer, he worked in optics in his spare time and developed an alternative to the dominant corpuscular theory of light.

Galilei, Galileo (1564–1642)   Astronomer, physicist, and philosopher who was interested in mechanics, optics and the motion and nature of "heavenly bodies." He argued against the principles of Aristotle's physics with both thought experiments and evidence he collected via actual experiments and observations aided by a telescope. Author of *Dialogue Concerning the Two Chief World Systems* (1632) for which he was condemned for heresy.

Giere, Ronald   Contemporary philosopher of science interested in scientific reasoning and explanation. Author of *Explaining Science: A Cognitive Approach* (1988).

Hacking, Ian   Contemporary philosopher of science who writes on conceptual change, social construction of reality and scientific realism. Author of *Representing and Intervening* (1983), where he discusses the role of experimentation in the practice of science, and of *The Social Construction of What?* (1999).

Harvey, William (1578–1657)   Very successful physician (he was the doctor of King James I and King Charles I), researcher, and lecturer (physiology and embryology). In his *Anatomical Essay on the Motion* *of the Heart and Blood in Animals* (1628), he offered a detailed description of the functioning of the circulatory system, where he explains the role of the heart as a pump. He based his work on empirical investigations conducted on the live bodies of animals and the dead bodies of humans.

Hawthorne effect   This is an effect that can skew the results of a study and negatively impact on the methodology of an experiment. In psychology, it is often used to account for how participants tend to behave in a way that pleases the experimenter or confirms the experimenter's expectations. Other similar effects are placebos and the Pygmalion effect.

holism (of meaning)   This is the view that linguistic expressions get their meanings in relation to other expressions in the system and the change of meaning of one expression determines a change of meaning in all the others as they are interrelated.

Hume, David (1711–1776)   Philosopher and historian who defended empiricism and provided a critical analysis of causal and inductive inferences (giving rise to the famous problem of induction). Author of *A Treatise of Human Nature* (1739–1740) and the *Enquiries concerning Human Understanding* (1748).

hyperbolic doubt   When Descartes wanted to find a truth that he could not doubt, in order to base on that truth his method for philosophical investigation, he imagined the existence of an evil demon with the constant aim of deceiving. Among other things, the existence of this demon would cause Descartes to doubt all of his sense experience. The hyperbolic doubt is often taken to be the starting point of skeptical arguments against naïve realism.

indexical   A linguistic expression whose reference changes according to the circumstances. The sentence "I am tired today" can have different meaning depending on who is uttering it and when it is uttered, because "I" and "today" are indexicals.

induction   Mode of inference in which the truth of the premises makes the conclusion probable but not necessarily true.

inductive statistical model   View of scientific explanation put forward by Hempel to account for the logical relation between an event to be explained and the factors that contribute to its explanation, when the event is the conclusion of an inductive argument which has the initial conditions and a highly probable generalization as its premises.

inductivism   Picture of scientific reasoning and practice according to which scientists arrive at explanatory theories by making observations and formulating generalizations on the basis of their specific observations.

inference to the best explanation   Mode of inference in which the conclusion is supported because it being true is the best explanation of a known event given all the available evidence.

internal realism   Form of moderate realism (endorsed by Putnam) according to which one can discover truths by scientific investigations, but these truths are internal (and to some extent relative) to a conceptual framework.

isomorphism From the Ancient Greek, the term means "sameness of form, of structure" and is used to describe relations, e.g. the relation between a model and a set of phenomena the model is supposed to represent; or the relation between a theory and the reality the theory is aimed at describing and explaining.

justification A process by which a belief receives support on the basis of good evidence or good argument.

Kepler, Johannes (1571–1630) Mathematician and astronomer interested both in optics and cosmology who is responsible for the view that planets move in elliptical orbits and for the three laws of planetary motion (now known as Kepler's Laws).

Kuhn, Thomas (1922–1996) Physicist, philosopher, and historian of science, Kuhn left his mark with the publication of *The Structure of Scientific Revolutions* (1st ed. 1962), in which he challenged the view that there is steady and cumulative progress in science, and introduced psychological and sociological factors in the explanation of scientific theory change.

Lakatos, Imre (1922–1974) Philosopher of science whose work was inspired by that of Karl Popper and who developed original ideas on the problem of demarcation and the best way to account for scientific change. In his most influential work, a 1970 paper entitled "The Methodology of Scientific Research Programmes," he defends the rationality of science and the notion of progress from the challenges made by Kuhn.

Laudan, Larry Contemporary philosopher of science who wrote extensively on progress and rationality in science. He has defended a sophisticated account of falsificationism and argued against relativism and subjectivism in philosophical accounts of scientific change. Author of *Progress and its Problems* (1977).

Lavoisier, Antoine (1743–1794) Chemist who showed the inconsistency of the phlogiston theory in *Reflections on Phlogiston* (1783) and was mainly responsible for the discovery of oxygen, giving rise to the chemical revolution. In *Elementary Treatise of Chemistry* (1789) he presented a unified account of the knowledge of chemistry of his time and praised the role of observations and experiments.

lawlike statements Lawlike statements are statements that, if true, express a law of nature. What the features of lawlike statements should be is, to some extent, open to debate.

laws of nature General principles that seem to play a central role within a scientific theory and in explanation and prediction of particular phenomena. Which defining characteristics a law has (whether it captures a necessary truth; whether it supports counterfactuals; whether it articulates a causal connection; etc.) is controversial.

Lipton, Peter (1954–2007)   Philosopher who contributed significantly to the epistemology (e.g. explanation, evidence, testimony) and the metaphysical issues (realism) in the philosophy of science. His book *Inference to the Best Explanation* (first published in 1991, second edn 2004) is a classic.

logical equivalence (principle of)   If two hypotheses are logically equivalent, then any observation statement that supports one of the hypotheses will support the other too.

logical form   Logical form is the structure of a proposition or an argument, exemplified by strings of symbols which obey syntactical rules of formation. The logical form of an argument is the pattern of inference which is obtained by abstracting from the content of its premises and conclusions.

logical positivism (or logical empiricism)   Philosophical movement originally founded in Vienna in 1922 by Moritz Schlick (a physicist), Otto Neurath (an economist), and Philipp Frank (a physics professor). Aimed at promoting and divulging a "scientific conception of the world," it was influenced by the philosophy of Bertrand Russell and Ludwig Wittgenstein. The "positivist" label refers to the view defended by these authors that scientific knowledge is the only legitimate type of knowledge and therefore science is especially valuable. The "empiricist" label derives from their view that knowledge about the world cannot be obtained without relying on experience, via observations and empirical tests.

logical terms   Logical terms are terms that stand for predicates or relations in a formal language. For instance, "P" in (x) Px→Qx, is a logical term standing for a predicate; "→" is a logical connective which stands for a conditional relation between Px and Qx (*if* x is P, *then* x is Q).

Lucretius (first century BC)   Latin poet who wrote *On the Nature of Things*, in which he argues for physical and cosmological views endorsed by the philosopher Epicurus (third century BC) and especially for atomism, the infinity of the universe and the mortality of the soul.

Mach, Ernst (1838–1916)   Physicist and philosopher who worked in optics and mechanics with impressive results which opened the way to relativity theory. In philosophy, he is known for the thesis that all knowledge comes from sensations and even laws of nature are a useful way to describe relations among sensation. This view is defended in *Contributions to the Analysis of the Sensations*, 1897.

Maxwell, James Clerk (1831–1879)   Mathematician and physicist known for discovering simple equations that express the relation between electric and magnetic fields and for defending the view that light is an electromagnetic phenomenon. Author of *Electricity and Magnetism* (1873).

mechanism   In general terms, this is the view according to which the behavior of a system (e.g. the Universe) can be explained by reference to material particles governed by deterministic laws. It is often contrasted with organicism.

metaphysics Study of what exists, of what is real. Branch of philosophy traditionally characterized by an attempt to disclose general principles and the ultimate nature of reality, over and beyond the observable properties of existing things and events that are the object of science.

methodological pluralism The view that science is not a unitary activity and its methods cannot be captured by a general account for natural and social sciences.

Mill, John Stuart (1806–1873) Philosopher and economist whose views are associated with a defense of empiricism and liberal values. He endorsed utilitarianism in ethics and wrote about methodology, identifying the fundamental principles of scientific practice in the *System of Logic, Ratiocinative and Inductive* (1843). Also author of *On Liberty* (1859).

model A model is a fiction, an object, structure or description that represents some features of a theory (e.g. the double helix model of DNA). What models are (more precisely) and what role they play in scientific discovery and explanation, is a controversial issue.

natural kinds Way of classifying things to be found in nature (as opposed to artifacts). For the essentialist, natural kinds group things together when they have the same essential properties (often identified with an underlying physical structure). For the anti-essentialist, natural kinds group things together when this grouping serves some useful purpose (e.g. for the purpose of making generalization or explaining phenomena).

naturalism In the context of the philosophy of social sciences, this is the view according to which there is continuity of aims and methods between social and natural sciences, and both types of science are aimed at discovering truths about natural and social facts by uncovering relevant causal connections.

necessary Opposite of "contingent." Something that is by necessity and could not have been otherwise. It is necessary that hexagons have six sides.

Newton, Isaac (1642–1727) Physicist and mathematician who wrote *Mathematical Principles of Natural Philosophy* (knows as *Principia*) published in 1687, where he states the foundations of mechanics and introduces the law of gravitation.

normal science Period characterized by the development of a scientific discipline within a certain paradigm.

normative A statement is normative if it expresses a judgment about what ought to be the case.

objective probability Probability of an event occurring given available data on observable events.

observation term A term is observational if it refers to observable objects or properties of objects. E.g. In the statement: "The autumn leaves are red," "red" is an observation term. More controversial is whether "temperature" is an observation term.

organicism   In general terms, this is the view according to which a system (e.g. the Universe) operates in the same way as a living organism with consciousness or intentions, and therefore escapes the predictability of cause and effect relations which govern all inanimate matter. Different or more specific characterizations of this term can be found in biology and theory of art.

paradigm   A framework combining the theoretical, methodological, and metaphysical assumptions which shape the work of the scientific community in periods of normal science. There is a huge literature on the different uses of the term "paradigm" which entered common discourse after the notion was explored by Kuhn in his account of scientific revolutions.

pathological science   Uses of the term "pathological" as applied to science differ considerably in the literature, but what seems to be a common structure in the examples of pathology in the natural sciences is that (1) some scientists hypothesize the existence of some entity in order to explain some phenomena; (2) such an entity can be inferred only as the result of a complex experimental technique which did not seem to be delivering the same results once adopted by other scientists or groups of scientists; (3) it turns out that the entity was fictitious. Apart from problems of replicability of experimental results, pathological examples of science are often characterized by unwarranted credulity in the scientific community and *ad hoc* defenses of the claims made by the "pathological" scientists.

Plato (427–347 BC)   Athenian philosopher who wrote many dialogues on issues within epistemology, ethics, politics and metaphysics using the historical figure of Socrates as the main character.

Poincaré, Henri (1854–1912)   Mathematician and physicist who wrote about the philosophy of science and especially about scientific discovery and confirmation. He argued for conventionalism, i.e. the view that theories are true by convention. Author of *Science and Hypothesis* (1902), *The Value of Science* (1905), and *Science and Method* (1908).

Popper, Karl (1902–1994)   Very influential philosopher of science who contributed greatly to the debate on the demarcation criterion, the confirmation of scientific theories and the nature of scientific change and progress. Although he was critical of many of the logical positivist views, he shared their conception of science as a paradigmatic achievement of human rationality. He defended falsificationism in *Conjectures and Refutations: The Growth of Scientific Knowledge* (1963).

posterior probability   Probability of an event/hypothesis after new evidence has been taken into account.

Priestley, Joseph (1733–1804)   Chemist who helped to develop and strenuously defended the phlogiston theory against emerging doubts. Author of *Experiments and Observations on Different Kinds of Air*, 1774.

prior probability Probability of an event/hypothesis before new evidence is taken into account.

projectible A predicate is projectible if we can expect it to apply in the future to the same object it has applied to so far. Goldman created a predicate that was not projectible, "grue."

pseudoscience A discipline or theory is pseudo-scientific if it has the superficial features of a scientific discipline or theory but fails to satisfy the accepted criteria for science. Common (but to some extent controversial) examples are astrology and creationism.

psychological approach to moral status The view that the psychological capacities of an individual are relevant in order to establish whether that individual has moral status.

puzzle-solving Activity that characterizes normal science for Kuhn. This is when scientists focus their efforts on explaining facts by using the resources provided by the dominant theory.

Quine, W. V. O. (1908–2000) Mathematical logician and very influential philosopher known for attacking the dichotomy between analytic and synthetic, developing naturalized epistemology and conceiving of philosophy as a branch of science. Responsible for a radical version of the so-called Duhem–Quine thesis and author of *Word and Object* (1960).

rationalism (adj. rationalist) In epistemology rationalism is the view that we can acquire knowledge independently of sense experience. In philosophy of science, the term can also be used to refer to those who think that scientific change obeys rational criteria of theory choice.

reference The relation between the linguistic expressions we use and the entities for which they stand. If I have a dog named Fido, the name "Fido" refers to my dog.

Reichenbach, Hans (1891–1953) Philosopher of science who belonged to the Logical Positivism movement, promoted the verifiability principle and studied the concept of space and time as they were represented in the relativity theory. Author of *The Rise of Scientific Philosophy* (1951).

relativism (conceptual) In philosophy of science, this is the view that scientists committed to different paradigms employ different concepts and have a different worldview as a consequence. What is true about Newtonian mass might not be true about Einsteinian mass and there is no neutral ground on which to adjudicate which notion of mass is the correct one independently of the commitment to a paradigm.

representational A statement is representational if its truth or falsehood depends on how things objectively are. Only representational statements have cognitive significance for the logical positivists.

research programs Units of scientific practice characterized by an evolving set of theories, similar methodology, and core theoretical principles. A research program is progressive if subsequent theories can predict novel facts, have more explanatory power and fit the evidence better than the previous ones. Otherwise, it is degenerating.

revolution   According to Kuhn, a revolution is a dramatic change of paradigm in the development of a scientific discipline characterized by a replacement of the dominant theory and by the introduction of new metaphysical and methodological assumptions. Philosophers who deny that there is ever dramatic change in science will deny the occurrence of scientific revolutions.

rigid designator   A rigid designator is a term that refers to the same object or property in all possible worlds and could not refer to anything else. If we think that "water" is a rigid designator, in all those worlds in which there is no substance with the same characteristics as water, the term "water" will not refer. What the defining or essential characteristics of water are is open to debate. Putnam believes that natural-kind terms are rigid designators and that "water" designates all substances with the same microstructure as what we call "water" on Earth (H_2O).

Russell, Bertrand (1872–1970)   Logician and philosopher who is considered one of the founders of analytic philosophy. He was interested in issues raised by scientific methodology and conceptual analysis. Author of *Principles of Mathematics* (1903) and *Problems of Philosophy* (1912), where he addresses the problem of induction. He also endorsed a descriptivist theory of reference in "On Denoting" (1905).

Salmon, Wesley (1925–2001)   Historian and philosopher of science who wrote extensively on the issues of theory confirmation and probability, induction, explanation and causality. Author of *The Foundations of Scientific Inference* (1967), and *Scientific Explanation and the Causal Structure of the World* (1984).

scientific method   The method by which scientists operate. This can be interpreted descriptively or normatively. Philosophers aiming at describing scientific method will ask how scientists happen to operate within their scientific communities – which type of reasoning they follow most of the time; how they reach certain conclusions; etc. Philosophers interested in the normative dimension will ask how scientists ought to operate for science to be progressive and contribute to knowledge. Opinions differ on whether it is possible to codify a general methodology for all scientific disciplines or whether we should settle for discipline-relative methodologies.

semantic view of theories   According to this view, which comes in different versions, theories should not be understood as interpretations of axiomatic systems and models play an important role in scientific understanding. This view is defended by Bas Van Fraassen and Ronald Giere.

semantics (adj. semantic)   Study of the meaning of linguistic expressions.

sense   Concepts associated by a particular speaker or a linguistic community to the use of a linguistic expression. The term "Earth" will be associated by a reasonably well-educated speaker today with the thought of a planet revolving around the Sun and inhabited by humans.

Sober, Elliott   Contemporary philosopher of science and of biology who has participated in debates on the distinction between science and pseudoscience, the role of simplicity in theory evaluation, and the roots of altruism. Author of *From a Biological Point of View* (1994) and *Reconstructing the Past* (1988), where he addresses the problem of induction.

sound   A deductive argument is sound if it is valid and has true premises.

Stahl, Georg (1659–1734)   Physician and chemist who developed the phlogiston theory, first to explain combustion, and then to account for a variety of chemical phenomena. Author of the *Opusculum Chymico-Physico-Medicum* (1715).

Strawson, Peter (1919–2006)   Philosopher interested in logic and in the project of developing a descriptive metaphysics. He argued for the dissolution of the problem of the justification of induction. Author of *Introduction to Logical Theory* (1952), *Individuals* (1959) and *The Bounds of Sense* (1966).

subjective probability   Judgment of probability based on the degree of belief that an agent has in the occurrence of a particular event.

Suppe, Frederick   Contemporary philosopher of science who developed the semantic view of scientific theories. Author of the *Semantic Conception of Theory and Scientific Realism* (1989) and editor of an influential collection of papers entitled *The Structure of Scientific Theories* (1977).

syntactic view of theories   According to the syntactic view, a scientific theory can be formalized in an axiomatic system which reveals the skeleton or structure of the theory. The main claim is that the interpretation of the theory, e.g. what the theory is about, can be separated from the structure of the formalized system. This view is defended by Carnap and Hempel. Opponents typically defend the semantic view of theories.

syntax (adj. syntactic)   Study of the rules that govern the structure of sentences.

synthetic   A statement is synthetic if its truth or falsehood depends neither on its logical structure nor on the meanings of the terms it contains. E.g. "All tall buildings are ugly"; "Cats are lazy."

testimony   Way of acquiring beliefs not by direct evidence or personal experience but on the basis of someone else's report. Testimony can be reliable or unreliable depending on the circumstances in which it is acquired.

Thagard, Paul   Philosopher of the cognitive sciences interested in scientific explanation and in conceptual change. Author of *Conceptual Revolutions* (1992).

theoretical definitions   A theoretical definition is a stipulation by which a new term is introduced within a theoretical framework (e.g. "quantum") or a previously used term is given a new connotation on the basis of a theory change (e.g. "mass"). The definition of a theoretical term should make clear what the role of the entity or property to which the term refers is within the theory.

theoretical terms   Terms which typically do not refer to observable entities or properties, but to the entities or properties posited by a scientific theory for explanatory and predictive purposes (e.g. "electron," "phlogiston," "conditioning," etc.) Whether theoretical terms genuinely refer and how their meaning is determined is the object of controversy – see the realism/anti-realism debate and the debate between descriptivists and causal theorists of reference.

theory-laden   An observation is theory-laden when theoretical assumptions affect the content of the observation statements that are derived from it. These theoretical assumptions can sometimes be necessary for developing instruments which aid observation, or can be embedded (often in a non-explicit way) in the language by which the observation is reported. The claim that observation is theory-laden is supposed to challenge the view that observation is neutral with respect to rival theoretical approaches.

theory/observation distinction   This is the idea that theoretical statements and observation statements differ in some significant way (e.g. their truth is determined via a different process; or all the terms contained in observation statements refer, whereas not all the terms contained in theoretical statements refer; etc.). The distinction is stressed by the logical positivists and is important for the plausibility of the verifiability principle and the classic conception of theories. The distinction is undermined by the sociologists of science who appeal to the theory-ladenness of observation statements and the claim that the world of our experience changes after a scientific revolution. The distinction also plays a crucial role in the realism/anti-realism debate.

thought experiments   Situations which are imagined and not real and whose outcomes are supposed to show something relevant to the truth of some theory, principle, or statement. The purpose of thought experiments and their methodological role in science and philosophy are controversial.

underdetermination   A set of hypotheses or theories are underdetermined by the available evidence if they are empirically equivalent (that is, evidence cannot discriminate between them) but incompatible (that is, they cannot both be true). Underdetermination comes in different strengths. Some claim that evidence can never be sufficient to determine theory choice (strong thesis) whereas others settle for the more modest claim that in some circumstances theory choice is underdetermined by evidence (weak underdetermination).

utilitarian calculus   Process by which we determine whether an action is right or wrong on the basis of its overall consequences for all the individuals involved. What is calculated is utility, which can be characterized in various ways (e.g. in terms of well-being, happiness, or satisfaction of relevant preferences/interests). Approach to ethics defended by Jeremy Bentham, John Stuart Mill and, more recently, Peter Singer.

valid   A deductive argument is valid if it is impossible for the premises to be true and the conclusion to be false.

Van Fraassen, Bas   Contemporary philosopher of science who developed constructive empiricism as an alternative to realism and instrumentalism in his influential book, *The Scientific Image* (1980). He also defends a pragmatic view of explanation and the theory/observation distinction.

verifiability   Criterion that a synthetic statement needs to satisfy in order to be regarded as meaningful by the logical positivists. A synthetic statement has meaning if we can determine its truth on the basis of some empirical evidence or generate predictions from it that are confirmed by evidence.

verification   Verification is the process by which a hypothesis or theory is tested. The hypothesis (or theory) is verified, that is, regarded as true, if the predictions made on its basis have so far been confirmed by evidence.

virtue-ethics   Approach to ethics that focuses on the importance of agents developing a good moral character rather than following rules or engaging in calculation of the possible effects of their actions. Aristotle is regarded as the philosopher mainly responsible for this approach to ethics.

Thematic Bibliography

General Readings on Philosophy of (the) Science(s)

Baert, P. (2005) *Philosophy of the Social Sciences*. Blackwell.

Baird, D., McIntyre, L. C., and Scerri, E. R. (eds) (2006) *Philosophy of Chemistry: Synthesis of a New Discipline*. Springer.

Bechtel, W. (ed.) (2001) *Philosophy and the Neurosciences*. Blackwell.

Bermudez, J. (2005) *Philosophy of Psychology*. Routledge.

Bird, A. (1998) *Philosophy of Science*. UCL Press.

Boyd, R. and Gasper, P. (eds) (1991) *The Philosophy of Science*. MIT Press.

Boyd, R., Gasper, P., and Trout, J. D. (eds) (1991) *The Philosophy of Science*. Blackwell.

Brody, B. A. and Grandy, R. E. (eds) (1989) *Readings in the Philosophy of Science* (2nd edn). Prentice-Hall.

Butterfield, J., Earman, J., Gabbay, D. M., Thagard, P., and Woods, J. (eds) (2006) *Philosophy of Physics*. Elsevier.

Cooper, R. (2007) *Psychiatry and Philosophy of Science*. Acumen.

Curd, M. and Cover, J. A. (1998) *Philosophy of Science: The Central Issues*. W. W. Norton and Co. Ltd.

DeWitt, R. (2004) *Worldviews: An Introduction to the History and Philosophy of Science*. Blackwell.

Gabbay, D. M., Thagard, P., and Woods, J. (eds) (2006) *Philosophy of Psychology and Cognitive Science*. Elsevier.

Godfrey-Smith, P. (2003) *Theory and Reality: An Introduction to the Philosophy of Science*. University of Chicago Press.

Harding, S. G. (ed.) (1976) *Can Theories be Refuted?: Essays on the Duhem–Quine Thesis*. Springer.

Hauman, D. (1994) *Philosophy of Economics* (2nd edn). Cambridge University Press.

Hitchcock, C. (ed.) (2004) *Contemporary Debates in Philosophy of Science*. Blackwell.

Kukla, A. (2001) *Methods of Theoretical Psychology*. MIT Press.

Ladyman, J. (2002) *Understanding Philosophy of Science*. Routledge.

Lange, M. (2002) *An Introduction to the Philosophy of Physics*. Blackwell.

— (ed.) (2007) *Philosophy of Science: An Anthology*. Blackwell.

Machamer, P. and Silberstein, M. (ed.) (2002) *The Blackwell Guide to the Philosophy of Science*. Blackwell.

Martin, M. and McIntyre, L. C. (eds) (1994) *Readings in the Philosophy of Social Science*. MIT Press.

Newton-Smith, W. H. (ed.) (2001) *A Companion to the Philosophy of Science*. Blackwell.

Nola, R. and Irzik, G. (2006) *Philosophy, Science, Education and Culture*. Springer.

Papineau, P. (ed.) (1996) *The Philosophy of Science*. Oxford University Press.

Psillos, S. (2007) *Philosophy of Science A–Z*. Edinburgh University Press.

Rosenberg, A. (2005) *Philosophy of Science: A Contemporary Introduction*. Routledge.

Rosenberg, A. and Macshea, D. (2007) *Philosophy of Biology: A Contemporary Introduction.* Routledge.

Sarkar S. and Pfeifer J. (2006) *Philosophy of Science: An Encyclopedia.* Taylor and Francis.

Sober, E. (1993) *Philosophy of Biology.* Oxford University Press.

Suppe, F. (ed.) (1977) *The Structure of Scientific Theories* (2nd edn). University of Illinois Press.

Thornton, T. (2007) *Essential Philosophy of Psychiatry.* Oxford University Press.

On Science, Pseudoscience, and Bad Science

Ayer, A. (1936, 2001) *Language, Truth and Logic* (chs 1, 2, and 6). Penguin Books.

Bauer, H. (1994) *Scientific Literacy and the Myth of the Scientific Method.* University of Illinois Press.

Bunge, M. (1984) "What is pseudo-science?" *Skeptical Inquirer* 9: 36–46.

Carnap, R. (1935) *Philosophy and Logical Syntax.* Kegan Paul.

Chalmers, A. (1999) *What is this Thing called Science?* (ch. 7) (3rd edn). University of Queensland Press.

Chauvin, R. (1999). "Psychological research and alleged stagnation." *Journal of Scientific Exploration* 13(2): 317–22.

Dupré, J. (1993) *The Disorder of Things: Metaphysical Foundations of the Disunity of Science.* Harvard University Press.

Eman, C. and Eman, M. (2002) "How not to be Lakatos intolerant." *International Studies Quarterly* 46: 231–62.

Feyerabend, P. (1979) *Science in a Free Society.* Routledge.

— (1975) *Against Method* (ch. 19). Verso Books.

French, S. (2007) *Science: Key Concepts in Philosophy.* Continuum.

Friedman, M. (1999) *Reconsidering Logical Positivism.* Cambridge University Press.

Fuller, S. (2007) *Science vs. Religion? Intelligent Design and the Problem of Evolution.* Polity Press.

Gratzer, W. B. (2000) *The Undergrowth of Science: Delusion, Self-Deception, and Human Frailty.* Oxford University Press.

Haack, S. (2005). "Trial and error: the supreme court's philosophy of science" *American Journal of Public Health* 95: S1–S66.

Hanfling, O. (ed.) (1981) *Essential Readings in Logical Positivism.* Blackwell.

Kitcher, P. (1982) *Abusing Science: The Case Against Creationism.* MIT Press.

Ladyman, J., Ross, D. with Spurrett, D., and Collier, J. (2007) *Every Thing Must Go: Metaphysics Naturalised.* Clarendon Press.

Lakatos, I. (1978) "Falsification and the methodology of scientific research programmes" in *The Methodology of Scientific Research Programmes: Philosophical Papers*, vol. 1. Cambridge University Press.

Langmuir, I. and Hall, R. N. (1989) "Pathological science" Physics Today 42 (10): 36–48.

Lansing, S. (2002) "'Artificial Societies' and the social sciences" *Artificial Life* 8, 279–82.

Laudan, L. (1983) "The demise of the demarcation problem" in Cohen, R. S. and Laudan, L. (eds) *Physics, Philosophy and Psychoanalysis*, pp. 111–27. Reidel.

Lewens, T. (2006) "Distinguishing treatment from research: a functional approach" *Journal of Medical Ethics* 32: 424–9.

Lilienfeld, S., Lynn, S. J., Lohr, J. (eds) (2004) *Science and Pseudoscience in Clinical Psychology.* Guilford Press.

Misak, C. J. (1995) *Verificationism: Its History and Prospects* (ch. 2). Routledge.

Park, R. (2000) *Voodoo Science: the Road from Foolishness to Fraud*. Oxford University Press.

Popper, K. (1963, 2002) *Conjectures and Refutations*. Routledge.

— (1959, 2002) *The Logic of Scientific Discovery*. Routledge. Appeared in German in 1934.

Reichenbach, H. (1951) *The Rise of Scientific Philosophy*. University of California Press.

Ruse, M. (1989) "Creation-science is not science" in Curd, M. and Cover, J. A. *Philosophy of Science: The Central Issues*. W. W. Norton and Co. Ltd.

Ruse, M. and Pennock, R. (1996) *But Is It Science?* Prometheus Books.

Schlick, M. (1938, 1979) "Form and content. An introduction to philosophical thinking" in M. Schlick (ed.) *Collected Papers*, vol. 1, 285–369. Reidel.

Thagard, P. (1978) "Why astrology is a pseudoscience" in Asquith, P. and Hacking, I. (eds) *Proceedings of the Philosophy of Science Association*, vol. 1. East Lansing, PSA.

On Natural vs. Social Sciences

Durkheim, E. (1938) *The Rules of Sociological Method*. Free Press.

Kincaid, H. (2004) "There are laws in the social sciences" in Hitchcock, C. (ed.) *Contemporary Debates in Philosophy of Science*, pp. 168–86. Blackwell.

— (1996) *Philosophical Foundations of the Social Sciences*. Cambridge University Press.

Kuhn, T. (1991). "The natural and the human sciences" in Hiley, D., Bohman, J., and Shusterman, R. (eds) *The Interpretive Turn*, pp. 17–24. Cornell University Press.

Papineau, D. (1978) *For Science in the Social Sciences*. Macmillan.

Popper, K. (1957, 2002) *The Poverty of Historicism*. Routledge.

Ryan, A. (ed.) (1973). *The Philosophy of Social Explanation*. Oxford University Press.

Taylor, C. (1985) *Philosophy and the Human Sciences*. Cambridge University Press.

— (1971) "Interpretation and the sciences of man" in *Monist* 25: 3–51.

Trigg, R. (1985) *Understanding Social Science*. Blackwell.

Weber, M. (1949) *The Methodology of the Social Sciences*. Macmillan.

Winch, P. (1972) *Ethics and Action*. Routledge.

— (1958, 1990) *The Idea of a Social Science and Its Relation to Philosophy*. Routledge and Kegan Paul.

On Induction and Inductivism

Armstrong, D. (1991) "What makes induction rational?" in *Dialogue* 30: 503–11.

Beebee, H. (2006) *Hume on Causation* (ch. 3). Routledge.

Black, M. (1954) *Problems of Analysis* (ch. 11). Routledge.

Hanson, N. R. (1958) *Patterns of Discovery*. Cambridge University Press.

Holland, J., Holyoak, K., Nisbett, R., and Thagard, P. (1986) *Induction: Processes of Inference, Learning, and Discovery* (chs 1, 2, and 11). MIT Press.

Howson, C. (2000) *Hume's Problem*. Oxford Clarendon Press.

Hume (1748, 2004) *An Enquiry Concerning Human Understanding*. Dover.

— (1739, 2000) *A Treatise of Human Nature*. Oxford University Press.

Lakatos, I. (ed.) (1968) *The Problem of Inductive Logic*. North Holland.

Lipton, P. (2004) *Inference to the Best Explanation* (2nd edn). Routledge.

Mill, J. S. (1843, 2002). *A System of Logic: Ratiocinative and Inductive*. University Press of the Pacific.

Nola, R. and Irzik, G. (2006) *Philosophy, Science, Education and Culture* (ch. 7). Springer.

Okasha, S. (2001) "What did Hume really show about induction?" *Philosophical Quarterly* 51: 307–28.

Papineau, D. (1992) "Reliabilism, induction & scepticism" *Philosophical Quarterly* 42: 1–20.

Popper, K. (1972, 1979) *Objective Knowledge: an Evolutionary Approach* (ch. 1). Oxford Clarendon Press.

— (1963, 2002) *Conjectures and Refutations.* Routledge.

— (1959, 2002) *The Logic of Scientific Discovery* (ch. 1). Routledge. Appeared in German in 1934.

— (1953, 1974) "The problem of induction" in Miller, D. (ed.) *Popper Selections.* Princeton University Press.

Rosenkrantz, R. (1971) "Inductivism and Probabilism" *Synthese* 23 (2–3): 167–205.

Russell, B. (1967, 2001) *The Problems of Philosophy.* Oxford University Press.

Salmon, W. (1974) "The pragmatic justification of induction" in Swinburne, R. (ed.) *The Justification of Induction*, pp. 85–97. Oxford University Press.

Skyrms, B. (1999) *Choice and Chance: an Introduction to Inductive Logic* (4th edn). University of California.

Sober, E. (2004) *Core Questions in Philosophy* (4th edn). University of Wisconsin.

— (1988) *Reconstructing the Past* (ch. 2). MIT Press.

Strawson, P. (1952) *Introduction to Logical Theory* (ch. 9). Methuen.

On Thought Experiments

Atkinson, D. (2003) "Experiments and thought experiments in natural science" in Galavotti, M. (ed.) *Observation and Experiment in the Natural and Social Sciences*, pp. 209–25. Boston Studies in the Philosophy of Science, vol. 232, Kluwer.

Bishop, M. (1999) "Why thought experiments are not arguments" *Philosophy of Science* 66 (4): 534–41.

Borsboom, D., Mellenbergh, G., and Van Heerden, J. (2002) "Functional thought experiments" *Synthese* 130: 379–87.

Brown, J. (2004) "Why thought experiments transcend empiricism" in Hitchcock, C. (ed.) *Contemporary Debates in Philosophy of Science*, pp. 24–43. Blackwell.

— (1991) *The Laboratory of the Mind* (chs 1 and 2). Routledge.

Gendler, T. (1998) "Galileo and the indispensability of scientific thought experiments" *British Journal for the Philosophy of Science* 49: 397–424.

Kuhn, T. (1977, 1979) *The Essential Tension: Selected Studies in Scientific Traditions and Change* (ch. 10). University of Chicago Press.

Norton, J. (2004) "Why thought experiments do not transcend empiricism" in Hitchcock, C. (ed.) *Contemporary Debates in Philosophy of Science*, pp. 44–66. Blackwell.

— (1996) "Are thought experiments just what you always thought?" *Canadian Journal of Philosophy*, 26 (3): 333–66.

Sorensen, R. (1992) *Thought Experiments.* Oxford University Press.

On the Nature of Scientific Theories and Models

Carnap, R. (1967) *Logical Structure of the World: Pseudoproblems in Philosophy.* University of California Press.

— (1966) *Philosophical Foundations of Physics.* Basic Books.

Frigg, R. and Hartmann, S. (2006) "Models in science" in *Stanford Encyclopedia of Philosophy*.http://plato.stanford.edu/entries/models-science/, accessed February 2008.

Giere, R. (2004) "How models are used to represent reality" *Philosophy of Science* 71 (suppl.): 742–52.

— (2000) "Theories" in Newton-Smith, W. (ed.) *A Companion to the Philosophy of Science*, pp. 515–24. Blackwell.

— (1988) *Explaining Science: a Cognitive Approach*. University of Chicago Press.

Hempel, C. (1970) "On the 'standard' conception of scientific theories" in Radner, M. and Winokur, S. (eds) *Minnesota Studies in the Philosophy of Science*, vol. 4, pp. 142–63. University of Minneapolis Press.

Mattingly, J. (2005) "The structure of scientific theory change" *Philosophy of Science* 72 (2): 365–89.

Suppe, F. (2002) *Representation and Invariance of Scientific Structures*. CSLI Publications.

— (1989) *The Semantic Conception of Theories and Scientific Realism*. University of Illinois Press.

Suppes, P. (1967) "What is a scientific theory?" in Morgenbesser, S. (ed.) *Philosophy of Science Today*, pp. 55–67. Basic Books.

Teller, Paul (2001) "Twilight of the perfect model" *Erkenntnis* 55, 393–415.

Thagard, P. (1993) *Computational Philosophy of Science*. MIT Press.

Thomson, P (2000), "Biology" in Newton-Smith, W. H. (ed.) *A Companion to the Philosophy of Science* (ch. 3). Blackwell.

— (1987) "A defence of the semantic conception of evolutionary theory" *Biology & Philosophy* 2(1): 26–32.

Van Fraassen, B. (1980) *The Scientific Image* (ch. 3). Oxford University Press.

On Confirmation and Probability

Carnap, R. (1950) *Logical Foundations of Probability*. University of Chicago Press.

Eells E. and B. Fitelson (2000) "Measuring confirmation and evidence" *Journal of Philosophy* XCVII: 663–72.

El-Gamal, M. A. and Grether, D. M. (1995) "Are people Bayesian? Uncovering behavioral strategies" *Journal of the American Statistical Association* 90 (432): 1137–45.

Gillies, D. (2000) *Philosophical Theories of Probability*. Routledge.

Glymour, C. (1980) *Theory and Evidence*. Princeton University.

Godfrey-Smith, P. (2003) "Goodman's problem and scientific methodology" *Journal of Philosophy* 100: 573–90.

Good, I. J. (1967) "The white shoe is a red herring" *British Journal for the Philosophy of Science* 17: 319–22.

Goodman, N. (1954, 2006) *Fact, Fiction and Forecast*. Harvard University Press.

Hacking, I. (2001) *An Introduction to Probability and Inductive Logic*. Cambridge University Press.

— (1994) "Entrenchment" in Stalker, D. (ed.) *Grue! The New Riddle of Induction*, pp. 193–224. Open Court.

— (1965) *The Logic of Statistical Inference*. Cambridge University Press.

Hempel, C. (2000) *Selected Philosophical Essays* (ch. 8). Cambridge University Press.

— (1945) "Studies in the logic of confirmation" *Mind* 54: 1–26. Re-printed in Hempel, C. (1965) *Aspects of Scientific Explanation*, pp. 3–46. Free Press.

Horwich, P. (1982) *Probability and Evidence*. Cambridge University Press.

Howson, C. (2000) "Evidence and confirmation" in Newton-Smith, W. H. (ed.) *A Companion to the Philosophy of Science* (ch. 17). Blackwell.

Howson, C. and Urbach, P. (eds) (2006), *Scientific Reasoning: the Bayesian Approach.* Open Court (3rd edn).

Jackson, F. (1994) "Grue" in Stalker, D. (ed.) *Grue! The New Riddle of Induction,* pp. 79–96. Open Court.

Jaynes, E. (2003) *Probability Theory: The Logic of Science.* Cambridge University Press.

Jeffrey, R. C. (2004) *Subjective Probability: The Real Thing.* Cambridge University Press

— (1983) *The Logic of Decision.* University of Chicago Press.

Kahneman, D., Slovic, P., and Tversky, A. (1982) *Judgment under Uncertainty: Heuristics and Biases.* Cambridge University Press.

Kelly, K. T. and Glymour, C. (2004) "Why probability does not capture the logic of scientific justification" in Hitchcock, C. (ed.) *Contemporary Debates in Philosophy of Science,* pp. 94–114. Blackwell.

Kyburg, H.E. (1974) *The Logical Foundations of Statistical Inference.* Reidel.

Lipton, P. (2007) "Ravens revisited" in O'Hear, A. (ed) *Royal Institute of Philosophy Supplement* 82: 75–95. Cambridge University Press.

Maher, P. (2004) "Probability captures the logic of scientific confirmation" in Hitchcock, C. (ed.) *Contemporary Debates in Philosophy of Science,* pp. 69–93. Blackwell.

Nassau, K. (1983) *The Physics and Chemistry of Color: The Fifteen Causes of Color.* John Wiley and Sons.

Nicod, J. (1923) "The logical problem of induction" (trans. Woods, M.) in Nicod, J. *Geometry and Induction.* University of California.

Salmon, W. C. (1990) "Rationality and objectivity in science" in Savage, C. W. (ed.) *Scientific Theories,* pp. 175–204. University of Minnesota Press.

— (1975) "Confirmation and relevance" in Maxwell, G. and Anderson. Jr, R. M. (eds) *Induction, Probability, and Confirmation. Minnesota Studies in the Philosophy of Science,* vol. 6, 3–36. University of Minnesota Press.

— (1971) *Statistical Explanation and Statistical Relevance.* University of Pittsburgh.

Strevens, M. (2003) *Bigger than Chaos: Understanding Complexity through Probability.* Harvard University Press.

Swinburne, R. G. (2002) (ed.) *Bayes's Theorem.* Oxford University Press.

— (1971) "The paradoxes of confirmation. A survey" *American Philosophical Quarterly* 8: 318–29.

Vranas, P. (2004) "Hempel's raven paradox: a lacuna in the standard Bayesian solution" *British Journal for the Philosophy of Science* 55: 545–60.

On Theory and Observation

Ayer, F. (1936) *Language Truth and Logic.* Dover.

Carnap, R. (1966) *Philosophical Foundations of Physics.* Basic Books.

Churchland, P. (1988) "Perceptual plasticity and theoretical neutrality: a reply to Jerry Fodor" *Philosophy of Science* 55, 167–87.

— (1979) *Scientific Realism and the Plasticity of Mind.* Cambridge University Press.

Feyerabend, P. (1981) "An attempt at a realistic interpretation of explanation" and "On the interpretation of scientific theories" in *Realism, Rationalism and Scientific Method* (Philosophical Papers, V.1). Cambridge University Press.

— (1975) *Against Method: Outline of an Anarchistic Theory of Knowledge.* New Left Books.

Fodor, J. (1991) "The dogma that didn't bark" *Mind* 100 (398): 201–20.

— (1984) "Observation reconsidered" *Philosophy of Science* 51: 23–43.

Gillies, D. (1993) *Philosophy of Science in the Twentieth Century* (ch. 7). Blackwell.

Hacking, I. (1983) *Representing and Intervening* (ch. 10). Cambridge University Press.

— (1981) "Do we see through a microscope?" *Pacific Philosophical Quarterly*, 62 (4): 305–22.

Hanson, N. (1965) *Patterns of Discovery*. Cambridge University Press.

Hempel, C. (1965) "Empiricist criteria of cognitive significance: problems and changes" in *Aspects of Scientific Explanation*, pp. 101–19. Free Press.

Kuhn, T. (1962, 1970) *The Structure of Scientific Revolutions* (2nd edn) (ch. 10). University of Chicago Press.

Maxwell, G. (1998) "The ontological status of theoretical entities" in Curd, M. and Cover, J. A. (eds) (1998) *Philosophy of Science: The Central Issues*, pp. 1052–63. W.W. Norton and Co.

Misak, C. J. (1995) *Verificationism, its History and Prospects* (ch. 2). Routledge.

Newton-Smith, W. H. (1981) *The Rationality of Science* (ch. 2). Routledge.

Popper, K. (1959, 2002) *The Logic of Scientific Discovery* (ch. 5 and app. 10). Routledge. Appeared in German in 1934.

Quine, W. V. O. (1993) "In Praise of Observation Statements" *Journal of Philosophy* 90, 3: 107–16.

Sellars, W. (1963) "The Language of Theories" in Sellars, W. *Science Perception and Reality*, pp. 118–26. Routledge.

Van Fraassen, B. (1980) *The Scientific Image* (chs 1 and 2). Oxford Clarendon.

Wright, C. (1986) "Realism, Observation and Verificationism" in Wright, C. (1993) *Realism, Meaning and Truth*. Blackwell.

On Explanation

Achinstein, P. (1983) *The Nature of Explanation*. Oxford University Press.

— (1981) "Can there be a model of explanation?" *Theory & Decision* 13, 3: 201–27.

Bromberger, S. (1966) "Why-questions" in Colodny R. (ed.) *Mind and Cosmos*, pp. 86–111. University of Pittsburg Press.

Cartwright, N. (1980) "The truth doesn't explain much" *American Philosophical Quarterly* 17 (2): 59–63.

Henderson, D. (2005) "Norms, invariance, and explanatory relevance" *Philosophy of the Social Sciences* 35: 324–38.

— (2002) "Norms, normative principles, and explanation" *Philosophy of Social Science* 32: 329–64.

Hon, G. and Rakover, S. (eds) (2001) *Explanation: Theoretical Approaches and Applications*. Springer.

Knowles, D. (ed.) (1990) *Explanation and its Limits*. Cambridge University Press.

Lewis, D. (1979) "Counterfactual dependence & time's arrow" *Noûs* 13: 455–76. Reprinted in Lewis, D. (1983) *Philosophical Papers*, vol. 2. Oxford University Press.

Psillos, S. (2002) *Causation and Explanation* (chs 8 and 9). Acumen.

Salmon, W. C. (1989) *Four Decades of Scientific Explanation*. University of Minnesota Press.

— (1977) "Why ask 'why?'" *Proceedings and Addresses of the American Philosophical Association* 51: 683–705.

Sosa, E. and Tooley, M. (eds) (1993) *Causation*. Oxford University Press.

Strawson, G. (1989) *The Secret Connexion: Causation, Realism & David Hume*. Clarendon Press.

Thagard, P. (1991) "Philosophical and computational models of explanation" in *Philosophical Studies* 64 (1): 87–104.

Thagard, P. and Litt, A. (2008) "Models of scientific explanation" in Sun. R (ed.) *Cambridge Handbook of Computational Psychology* (ch. 20). Cambridge University Press.

Wilson, F. (1985) *Explanation, Causation and Deduction*. Kluwer.

Woodward, J. (2000) "Explanation and invariance in the special sciences" *British Journal for the Philosophy of Science* 51: 197–254.

— (1979) "Scientific explanation" *British Journal for the Philosophy of Science* 30: 41–67.

On Laws

Armstrong, D. (1983) *What is a Law of Nature*. Cambridge University Press.

Beebee, H. (2000) "The nongoverning conception of laws of nature" *Philosophy and Phenomenological Research* 61: 571–94.

Bigelow, J., Ellis, B., and Lierse, C. (1992) "The world as one of a kind: natural necessity and laws of nature" *British Journal for the Philosophy of Science* 43: 371–88.

Bird, A. (2005) "The dispositionalist conception of laws" *Foundations of Science* 10: 353–70.

Carroll, J. (1994) *Laws of Nature*. Cambridge University Press.

Cartwright, N. (1983) *How the Laws of Physics Lie*. Oxford University Press.

— (1980) "Do the laws of physics state the facts" *Pacific Philosophical Quarterly* 61: 75–84.

Dretske, F. (1977) "Laws of nature" *Philosophy of Science* 44: 248–68.

Earman, J. (1978) "The universality of laws" *Philosophy of Science* 45: 173–81.

Ellis, B. (2001) *Scientific Essentialism*. Cambridge University Press.

Giere, R. (1999) *Science without Laws*. University of Chicago Press.

Lange, M. (2000) *Natural Laws in Scientific Practice*. Oxford University Press.

Lewis, D. (1973) *Counterfactuals*. Harvard University Press.

Lipton, P. (1999) "All else being equal" *Philosophy* 74: 155–68.

Mumford, S. (2004) *Laws in Nature*. Routledge.

Swoyer, C. (1982) "The nature of natural laws" *Australasian Journal of Philosophy* 60: 203–23.

Tooley, M. (1977) "The nature of laws" *Canadian Journal of Philosophy* 7: 667–98.

Van Fraassen, B. (1989) *Laws and Symmetry*. Clarendon Press.

On Meaning, Reference, and Natural Kinds

Bird, A. (2007) *Nature's Metaphysics: Laws and Properties*. Oxford University Press.

Boyd, R. (1991) "Realism, anti-foundationalism and the enthusiasm for natural kinds" *Philosophical Studies* 61: 127–48.

Devitt, M. (1990) "Meanings just ain't in the head" in Boolos, G. (ed.) *Meaning and method: Essays in honor of Hilary Putnam*, pp. 79–104. Cambridge University Press.

— (1981) *Designation*. Columbia University Press.

Dupré, J. (1981) "Natural kinds and biological taxa" *Philosophical Review* 90: 66–90.

Evans, G. (1982) *The Varieties of Reference*. Oxford University Press.

— (1973) "The causal theory of names" in *Proceedings of the Aristotelian Society*, suppl. vol. 47: 187–208.

Frege, G. (1892, 1984) "On sense and meaning" in *Collected Papers on Mathematics, Logic and Philosophy*, pp. 157–77 (transl. by Black, M., Dudman, V., Geach, P., Kaal, H., Kluge, E., McGuinness, B., and Stoothoff, R.). Blackwell.

Hacking, I. (1992) "World-making by kind-making: child abuse for example" in Douglas, M. and Hull, D. (eds) *How Classification Works*, pp. 180–238. Edinburgh University Press.

— (1991) "A tradition of natural kinds" *Philosophical Studies* 91: 109–26.

— (1990) "Natural kinds" in Barrett, R. and Gibson, R. (eds) *Perspectives on Quine*, pp. 129–43. Blackwell.

Kripke, S. (1980) *Naming and Necessity*. Harvard University Press.

Kuhn, T. (1990) "Dubbing and redubbing: the vulnerability of rigid designation" in Savage, C.W. (ed.) *Scientific Theories. Minnesota Studies in the Philosophy of Science*, vol. 14, pp. 298–318. University of Minnesota Press.

LaPorte, J. (2003) *Natural Kinds and Conceptual Change*. Cambridge University Press.

Levin, M. (1979) "On theory-change and meaning-change" *Philosophy of Science* 46: 407–24.

Lewis, D. (1970) "How to define theoretical terms" *Journal of Philosophy of Science* 67: 427–46.

Mackie, P. (2007) *How Things Might Have Been*. Oxford University Press.

Mellor, H. (1977) "Natural kinds" *British Journal for the Philosophy of Science* 28: 299–312.

Miller, R. W. (2000) "Half-naturalized social kinds" *Philosophy of Science* 67: 640–52.

Papineau, D. (1996) "Theory-dependent terms" *Philosophy of Science* 63: 1–20.

Pessin, A. and Goldberg, S. (eds) (1996) *The Twin Earth Chronicles: Twenty Years of Reflection on Hilary Putnam's "The Meaning of 'Meaning'."* M. E. Sharpe.

Putnam, H. (1990) "Is water necessarily H_2O?" in Conant, J. (ed.) *Realism with a Human Face*, pp. 54–79. Harvard University Press.

— (1975) "The meaning of 'meaning'" in Putnam, H. (1975) *Mind, Language and Reality: Philosophical Papers*, vol. 2, pp. 215–71. Cambridge University Press.

— (1973) "Meaning and Reference" *Journal of Philosophy* 70: 699–711.

Quine, W.V.O. (1969): "Natural kinds" in Quine, W.V.O. *Ontological Relativity and Other Essays*, pp. 114–38. Columbia University Press.

Russell, B. (1905) "On denoting" *Mind* 14: 479–93.

Salmon, N. (1982) *Reference and Essence*. Princeton University Press.

Schwartz, S. P. (1977) *Naming, Necessity, and Natural Kinds*. Cornell University Press.

Searle, J. (1983) *Intentionality* (ch. 8). Cambridge University Press.

— (1969) *Speech Acts: An Essay in the Philosophy of Language*. Cambridge University Press.

Soames, S. (2002) *Beyond Rigidity: The Unfinished Semantic Agenda of Naming and Necessity*. Oxford University Press.

Strawson, P. (1957) "Proper names" *Proceedings of the Aristotelian Society*, suppl. vol. 31: 191–228.

Wilkerson, T. E. (1995) *Natural Kinds*. Avebury.

Zammito, J. H. (2004) *A Nice Derangement of Epistemes*. University of Chicago Press.

Zemach, E. (1976) "Putnam's theory of reference of substance terms" *Journal of Philosophy* 73: 116 – 27.

On "Species"

Dupré, J. (1995) *The Disorder of Things*. Harvard University Press.

Griffiths, P. (1999) "Squaring the circle: natural kinds with historical essences" in Wilson, R. (ed.) *Species: New Interdisciplinary Studies*, pp. 209–28. MIT Press.

Hull, D. (1988) *Science as a Process*. University of Chicago Press.

— (1965) "The effect of essentialism on taxonomy: two thousand years of stasis" *British Journal for the Philosophy of Science* 15: 314–26 and 16: 1–18.

Kitcher, P. (1989) "Some puzzles about species" in Ruse, M. (ed.) *What the Philosophy of Biology Is*, pp. 183–208. Kluwer.

— (1984) "Species" *Philosophy of Science* 51: 308–33.

Levine, A. (2001) "Individualism, type specimens, and the scrutability of species membership" *Biology and Philosophy* 15: 325–38.

Mayr, E. (1987) "The ontological status of species: scientific progress and philosophical terminology" *Biology and Philosophy* 2: 145–66.

Rosenberg, A. (1987) "Why does the nature of species matter?" *Biology and Philosophy* 2: 192–7.

Ruse, M. (1987) "Biological species: natural kinds, individuals, or what?" *British Journal for the Philosophy of Science* 38: 225–42.

Sober, E. (1984) *The Nature of Selection: Evolutionary Theory in Philosophical Focus*. MIT Press.

Wilkerson, T. E. (1993) "Species, essences and the names of natural kinds" *The Philosophical Quarterly* 43: 1–19.

Wilson, J. (1999) *Biological Individuality: The Identity and Persistence of Living Entities*. Cambridge University Press.

On "Jade"

Braddon-Mitchell, D. (2005) "Conceptual change and the meaning of natural kind terms" *Biology and Philosophy* 20: 859–68.

Jackson, F. (1998) *From Metaphysics to Ethics: A Defence of Conceptual Analysis*. Clarendon Press.

Kim, J. (1993) *Supervenience and Mind, Selected Philosophical Essays*. Cambridge University Press.

LaPorte, J. (2004) *Natural Kinds and Conceptual Change*. Cambridge University Press.

Mariam, D. (1997) "Kim's functionalism" *Noûs* 31: 133–48.

Platts, M. (1983) "Explanatory kinds" *British Journal for the Philosophy of Science* 34: 133–48.

Poncinie, L. (1985) "Meaning change for natural kind terms" *Noûs* 19: 415–27.

Stenwall, R. (2005) "Aspect kinds" in Persson, J. and Ylikoski, P. (eds) *Rethinking Explanation*, 193–203. Springer.

On Incommensurability

Agazzi, E. (1985) "Commensurability, incommensurability, and cumulativity in scientific knowledge" *Erkenntnis* 22: 51–77.

Barnes, B. and Bloor, D. (1982) "Relativism, rationalism and the sociology of knowledge" in Hollis, M. and Lukes, S. (eds) *Rationality and Relativism*, pp. 21–47. MIT Press.

Bird, A. (2000) *Thomas Kuhn*. Acumen.

Chen, X. (1990) "Local incommensurability and communicability" in Fine, A., Forbes, M., and Wessels, L. (eds) *Proceedings of the 1990 Biennial Meeting of the Philosophy of Science Association*, pp. 67–76. PSA.

Davidson, D. (1974) "The very idea of a conceptual scheme" *Proceedings & Addresses of the American Philosophical Association* 47: 5–20.

Devitt, M. (1979) "Against incommensurability" *Australasian Journal of Philosophy* 57: 29–50.

Duhem, P. (1906/1914, 1991) *The Aim and Structure of Physical Theory*. Princeton University Press.

Field, H. (1973) "Theory change and the indeterminacy of reference" *Journal of Philosophy* 70: 462–81.

Fine, A. (1975) "How to Compare Theories: Reference and Change" *Noûs* 9: 17–32.

Hacking, I. (1982) "Language, truth and reason" in Hollis, M. and Lukes, S. (eds) *Rationality and Relativism*, pp. 48 – 66. MIT Press.

Kitcher, P. (1978) "Theories, theorists and conceptual change" *Philosophical Review* 87: 519–47.

Poincaré, H. (1902, 2003) *Science and Hypothesis*. Dover.

Sankey, H. (1991). "Translation failure between theories" *Studies in History and Philosophy of Science* 22: 223–36.

Sellars, W. (1956) "Empiricism and the philosophy of mind" in Feigl, H. and Scriven, M. (eds) *Minnesota Studies in the Philosophy of Science*, vol. 1. University of Minnesota Press.

On Realism

Bird, A. (1998) *Philosophy of Science* (ch. 4). Routledge.

Boyd, R. (1990) "Realism, approximate truth, and philosophical method" in Savage, C.W. (ed.) *Scientific Theories, Minnesota Studies in the Philosophy of Science*, vol. 14, pp. 355–91. University of Minnesota Press.

Brown, J. R. (1994) "Explaining the success of science," in Brown, J. R. *Smoke and Mirrors: How Science Reflects Reality* (ch. 1). Routledge.

Churchland, P. (1982) "The anti-realist epistemology of Van Fraassen's *The Scientific Image*" *Pacific Philosophical Quarterly* 63: 226–35.

Churchland, P. and Hooker, C. A. (eds) (1985) *Images of Science: Essays on Realism and Empiricism*. University of Chicago Press.

Devitt, M. (1997) *Realism and Truth*. Princeton University Press.

Fine, A. (1984) "And not anti-realism either" *Noûs* 18: 51–65.

— (1984) "Natural ontological attitude" in Leplin, J. (ed.) *Scientific Realism*, pp. 83–107. University of California Press.

Goldman, A. (2007) "The underdetermination argument for brain-in-the-vat scepticism" *Analysis* 67 (1): 32–6.

Goodman, N. (1978, 1988) *Ways of Worldmaking*. Hackett.

Hacking, I. (1981) "Experimentation and scientific realism" *Philosophical Topics* 13: 71–88.

Horwich, P. (1991) "On the nature and norms of theoretical commitment" *Philosophy of Science* 58: 1–14.

— (1982) "Three forms of realism" *Synthese* 51 (2): 181–201.

Kuipers, T. (2000) *From Instrumentalism to Constructive Realism*. Springer.

Lepin, J. (ed.) (1985) *Scientific Realism*. University of California Press.

Mach, E. (1893, 1960) *The Science of Mechanics*, 6th edn, trans. McCormack, T. J. Open Court.

Maxwell, G. (1962) "The ontological status of theoretical entities" in Feigl, H. and Maxwell, G. (eds) *Minnesota Studies in the Philosophy of Science*, vol. 3. Minnesota Press.

Papineau, D. (1996) "Introduction" in Papineau, D. (ed.) *The Philosophy of Science*, pp. 1–20. Oxford University Press.

Pritchard, D. (2005) *Epistemic Luck*. Oxford Clarendon Press.

Psillos, S. (1999) *Scientific Realism: How Science Tracks Truth*. Routledge.

Putnam, H. (1987) *The Many Faces of Realism*. Open Court.

— (1981, 1999) "Brains in a vat" in Putnam, H. *Reason, Truth, and History*, Cambridge University Press. Reprinted in DeRose, K. and Warfield, T. A. (eds) (1999) *Skepticism: A Contemporary Reader*. Oxford University Press.

Rosen, G. (1994) "What is constructive empiricism?" *Philosophical Studies* 74: 143–78.

Van Fraassen, B. (1980) *The Scientific Image* (ch. 2). Oxford Clarendon Press.

Worrall, J. (1989) "Structural realism: the best of both worlds?" *Dialectica* 43 (1–2): 99–124.

Worrall, J. (1984) "An unreal image" *British Journal for the Philosophy of Science* 35: 65–80.

— (ed.) (1994) *The Ontology of Science*. Dartmouth Publishing.

On Underdetermination and the Pessimistic Meta-Induction

Boyd, R. (1973) "Realism, underdetermination, and a causal theory of evidence" *Noûs* 7(1): 1–12.

Devitt, M. (2005) "Scientific realism" in Jackson, F. and Smith, M. (eds) *The Oxford Handbook of Contemporary Philosophy* (ch. 26). Oxford University Press.

Duhem, P. (1969) *To Save the Phenomena*. University of Chicago Press.

Gillies, D. (1998) "The Duhem thesis and the Quine thesis" in Curd, M. and Cover, J. A. (eds) *Philosophy of Science: The Central Issues*, pp. 302–19. W.W. Norton and Co.

Horwich, P. (1982) "How to choose between empirically indistinguishable theories" *Journal of Philosophy* 79(2), 61–77.

Kitcher, P. (1993) *The Advancement of Science*. Oxford University Press.

Kukla, A. (2001). Theoreticity, underdetermination, and the disregard for bizarre scientific hypotheses. *Philosophy of Science* 68: 21–35.

— (1993) "Laudan, Leplin, empirical equivalence, and underdetermination" *Analysis* 53: 1–7.

Lange, M. (2002) "Baseball, pessimistic inductions and the turnover fallacy" *Analysis* 62: 281–5.

Laudan, L. (1981) "A Confutation of convergent realism" *Philosophy of Science*, 48: 19–49.

Laudan, L. and Leplin, J. (1993) "Determination underdeterred: reply to Kukla" *Analysis* 53: 8–16.

— (1991) "Empirical equivalence and underdetermination" *Journal of Philosophy* 88: 449–72.

Leplin, J. (1997) *A Novel Defence of Scientific Realism*. Oxford University Press.

Lewis, P. (2001) "Why the pessimistic induction is a fallacy" *Synthese* 129: 371–80.

Okasha, S. (2002) "Underdetermination, holism and the theory/data distinction" *Philosophical Quarterly* 52(208), 303–19.

— (2000) "The underdetermination of theory by data and the 'strong programme' in the sociology of knowledge" *International Studies in the Philosophy of Science* 14(3): 283–97.

Povinelli, D. and Vonk, J. (2004) "We don't need a microscope to explore the chimpanzee's mind" *Mind & Language* 19 (1): 1–28.

Psillos S. (1999) *Scientific Realism: How Science Tracks Truth*. Routledge.

— (1996) "Scientific realism and the 'pessimistic induction'" *Philosophy of Science* 63: 306–14.

Putnam, H. (1983) "Realism and reason" *Philosophical Papers* (section iii). Cambridge University Press.

— (1978) *Meaning and the Moral Sciences*. Routledge.

Quine, W.V.O. (1951) "Two dogmas of empiricism" *Philosophical Review* 60 (1): 20–43.

Saatsi, J. (2005) "Pessimistic induction and two fallacies" *Philosophy of Science* 72 (5): 1088–98.

Van Fraassen, B. (1980) *The Scientific Image*. Oxford Clarendon Press.

Worrall, J. (1982) "Scientific realism and scientific change" *Philosophical Quarterly*, 32(128): 201–31.

On Revolutions and Progress

Andersen, H., Barker, P., and Chen X. (2006) *The Cognitive Structure of Scientific Revolutions*. Cambridge University Press.

Aronson, J. L., Harré, R., and Way, E. C. (1994) *Realism Rescued: How Scientific Progress is Possible*. Duckworth.

Barnes, B. and Bloor, D. (1982) "Relativism, rationalism and the sociology of knowledge" in Hollis, M. and Lukes, S. (eds) *Rationality and Relativism*, pp. 21–47. MIT Press.

Bird, A. (2007) "What is scientific progress?" *Noûs* 41 (1): 64–89.

Chang, H. (2007). "Scientific progress: beyond foundationalism and coherentism." *Royal Institute of Philosophy Supplement* 82: 1–20.

Clow, A. and Clow, N. L. (eds) (1992) *The Chemical Revolution*. Gordon and Breach Science Publishers.

Cohen, B. (1980) *The Newtonian Revolution*. Cambridge University Press.

Conant, J. B. (1950) *The Overthrow of the Phlogiston Theory: The Chemical Revolution of 1775–1789*. Harvard University Press.

Dilworth, C. (2008). *Scientific Progress: A Study Concerning the Nature of the Relation between Successive Scientific Theories* (3rd rev. edn). Springer.

Gavroglu, K., Goudaroulis, Y., and Nicolacopoulos, P. (eds) (1989). *Imre Lakatos and Theories of Scientific Change*. Kluwer.

Hacking, I. (1993) "Working in a new world: the taxonomic solution" in Horwich, P. (ed.) *World Changes. Thomas Kuhn and the Nature of Science*, pp. 275–310. MIT Press.

— (1982) "Language, truth and reason" in Hollis, M. and Lukes, S. (eds) *Rationality and Relativism*, pp. 48–66. MIT Press.

— (1981) (ed.) *Scientific Revolutions*. Oxford University Press.

Hall, R. (1983) *The Revolution in Science 1500–1750*. Longman.

Harré, R. (ed.) (1975). *Problems of Scientific Revolutions: Progress and Obstacles to Progress in the Sciences*. Oxford University Press.

Henry, J. (1997) *The Scientific Revolution and the Origins of Modern Science*. Palgrave MacMillan.

Hooykaas, R. (1987) "The rise of modern science: when and why" *British Journal of the History of Science* 20: 453–73.

Hoyningen-Huene, P. (1993) *Reconstructing Scientific Revolutions: Thomas S. Kuhn's Philosophy of Science*. (transl. by Levine, A.). University of Chicago Press.

Katz, J. (1989) "Rational common ground in the sociology of knowledge" *Philosophy of the Social Sciences* 19: 257–71.

Kuhn, T. (1977, 1979) *The Essential Tension: Selected Studies in Scientific Traditions and Change* (ch. 13). University of Chicago Press.

— (1974) "Second thoughts on paradigms" in Suppe, F. (ed.) *The Structure of Scientific Theories*, pp. 459–82. University of Illinois Press.

— (1957, 1990) *The Copernican Revolution: Planetary Astronomy in the Development of Western Thought*. Harvard University Press.

Lakatos, I. (1978) *The Methodology of Scientific Research Programmes*. Cambridge University Press.

— (1970) "Falsification and the methodology of scientific research programmes" from Lakatos, I. and Musgrave, A. (eds) *Criticism and the Growth of Knowledge*. Cambridge University Press.

Lakatos, I. and Musgrave, A. (1970) (eds) *Criticism and the Growth of Knowledge.* Cambridge University Press.

— (1990) *Science and Relativism.* University of Chicago Press.

— (1987) "Progress or rationality? The prospects for normative naturalism" *American Philosophical Quarterly* 24(1): 19–31.

— (1984) "Dissecting the holist picture of scientific change" in Laudan, L. *Science and Values* (ch. 4). University of California Press.

— (1977) *Progress and its Problems: Towards a Theory of Scientific Growth.* University of California Press.

Longino, H. (1990) *Science as Social Knowledge: Values and Objectivity in Scientific Enquiry.* Princeton University Press.

Lohmann, S. (2004) "A Toy model of scientific progress" *American Journal of Economics and Sociology* 63 (1): 167–81.

Losee, J. (2003). *Theories of Scientific Progress: An Introduction.* Routledge.

Newton-Smith, W. H. (1981) *The Rationality of Science* (chs 3 and 5). Routledge.

Niiniluoto, I. and Tuomela, R. (eds) (1979). *The Logic and Epistemology of Scientific Change.* Acta Philosophica Fennica 30.

Popper, K. (1994) *The Myth of the Framework: In Defence of Science and Rationality* (chs 1, 2, and 3). Routledge.

— (1975) "The rationality of scientific revolutions" in Harré, R. (ed.) *Problems of Scientific Revolution: Progress and Obstacles to Progress in the Sciences*, pp. 72–101. Clarendon Press.

— (1963, 2002). *Conjectures and Refutations* (ch. 10). Routledge.

— (1959, 2002) *The Logic of Scientific Discovery.* Routledge. Appeared in German in 1934.

Radnitzky, G. and Andersson, G. (eds) 1978. *Progress and Rationality in Science.* Reidel.

Shapere, D. (1989) "Evolution and continuity in scientific change" *Philosophy of Science* 56: 419–37.

Shapin, S. (1984) "Pump and circumstance: Robert Boyle's literary technology" in Collins, H., Pinch, T., and Shapin, S. (eds) *Social Studies of Science*, pp. 481–520. Sage.

Rescher, N. (1978) *Scientific Progress.* Blackwell.

Shea, W. (1988) *Revolutions in Science: Their Meaning and Relevance.* Science History Publications.

Thagard, P. (1992) "The conceptual structure of the chemical revolution" *Philosophy of Science* 57 (2): 183–209.

— (1992) *Conceptual Revolutions.* Princeton University Press.

Titchener, E. (1912) "The schema of introspection" *American Journal of Psychology* 23: 485–508.

Trigg, R. (1993) *Rationality and Science.* Blackwell.

On Science and Society and Research Ethics

Council for International Organizations of Medical Sciences (CIOMS) (2002) *International Ethical Guidelines for Biomedical Research Involving Human Subjects.* www.cioms.ch/frame_guidelines_nov_2002.htm [accessed March 2008].

Feyerabend, P. (1978) *Science in a Free Society.* New Left Books.

Fuller, S. (1993) *Philosophy, Rhetoric and the End of Knowledge.* University of Wisconsin Press.

Harris, J. (2005) "Scientific research is a moral duty" *Journal of Medical Ethics* 31: 242–8.

Kitcher, P. (2001) *Science, Truth and Democracy*. Oxford University Press.

Kurtz, P. (ed.) (2007) *Science and Ethics: Can Science Help Us Make Wise Moral Judgments?* Prometheus Books.

Longino, H. (1990) *Science as Social Knowledge: Values and Objectivity in Scientific Inquiry*. Princeton University Press.

McMullin, E. (2000) "Values in science" in Newton-Smith, W. (ed) *Companion to the Philosophy of Science*. Blackwell.

Parsons, K. (2003) *The Science Wars: Debating Scientific Knowledge and Technology*. Prometheus Books.

Potter, E. (2006) *Feminism and Philosophy of Science: an introduction*. Routledge.

Radnitzky, G. and Bartley, W. (eds) (1987) *Evolutionary Epistemology, Rationality, and the Sociology of Knowledge*. Open Court.

Sorell, T. (1991) *Scientism* (chs 1 and 4). Routledge.

World Medical Association (WMA) (2004), *Declaration of Helsinki*.www.wma.net/e/policy/b3.htm. [accessed March 2008].

On Enhancement, Disability, and the Ethics of Genetic Engineering

Agar, N. (2004) *Liberal Eugenics: In Defence of Human Enhancement*. Blackwell.

Berry, R. (2007) *The Ethics of Genetic Engineering*. Routledge.

Burley, J. and Harris, J. (eds) (2004) *A Companion to Genethics*. Blackwell.

Daniels, N. (forthcoming) "Can anyone really be talking about ethically modifying human nature?" in Sevalescu, J. and Bostrom N. *Enhancing Humans*. Oxford University Press.

— (1985) *Just Health Care*. Cambridge University Press.

Farrell, C. (2004) "The genetic difference principle" *American Journal of Bioethics* 4(2): 21–8.

Feinberg, J. (1984) "Harming as wronging" in Feinberg, J. *Harm to Others* (ch. 3). Oxford University Press.

Harris, J. (1995) "Should we attempt to eradicate disability?" *Public Understanding of Science* 4(3): 233–42.

— (1993) "Is gene therapy a form of eugenics?" *Bioethics* 7 (2–3): 178–87.

— (1992) *Wonderwoman & Superman: Ethics & Human Biotechnology*. Oxford University Press.

Harris, J. and Holm, S. (2002) "Extended lifespan and the paradox of precaution" *Journal of Medicine and Philosophy* 27(3): 355–69.

Häyry, M. (2004) "There is a difference between selecting a deaf embryo and deafening a hearing child" *Journal of Medical Ethics* 30: 510–12.

Koch, T. (2001) "Disability and difference: balancing social and physical constructions" *Journal of Medical Ethics* 27: 370–6.

McKie, J. et al. (1998) *The Allocation of Health Care Resources*. Ashgate.

Reindal, S. M. (2000) "Disability, gene therapy and eugenics – a challenge to John Harris" *Journal of Medical Ethics* 26: 89–94.

Schichor, N., Simonet, J., and Canano, C. (2003) "Should we allow genetic engineering? A public policy analysis of germline enhancement" in Gilbert, S. and Zackin, E. (eds) *Developmental Biology Online* (ch. 21).

On the Ethics of Scientific Research involving Animals

American Psychological Association Board of Scientific Affairs Committee on Animal Research and Ethics (2007) *Guidelines for Ethical Conduct in the Care and Use of Animals*. www.apa.org/science/anguide.html [accessed March 2008].

Armstrong, S. and Botzler, R. (2003) *The Animal Ethics Reader.* Routledge.

Dupré, J. (2006) *Humans and Other Animals.* Oxford University Press.

Kant, I. (1785, 1998) *Groundwork of the Metaphysics of Morals* (edited by Gregor, M. J.). Cambridge University Press.

Linzey, A. and Clarke, P. (eds) *Animal Rights: A Historical Anthology.* Columbia University Press.

Nuffield Council on Bioethics (2005) *The Ethics of Research Involving Animals.* www.nuffieldbioethics.org/go/browseablepublications/ethicsofresearchanimals/report_230.html [accessed March 2008].

Regan, T. (1983) *The Case for Animal Rights.* University of California Press.

Royal Society (2006) *The Use of Non-Human Primates in Research* (or The Weatherall Report). http://royalsociety.org/downloaddoc.asp?id 5 3696 [accessed March 2008].

Singer, P. (1974, 2001) *Animal Liberation.* Harper.

On Deception in Psychological Research

American Psychological Association (2002). *Ethical Principles of Psychologists and Code of Conduct.* www.apa.org/ethics/ [accessed March 2008].

Bok, S. (1999) *Lying: Moral Choice in Public and Private Life.* Vintage.

British Psychological Society (1992) *Ethical Principles for conducting Research with Human Participants.* www.bps.org.uk/the-society/ethics-rules-charter-code-of-conduct [accessed March 2008].

Clarke, S. (1999) "Justifying deception in social science research" *Journal of Applied Philosophy* 16(2): 151–66.

Darley, J. M. and Batson, C. D. (1973) "From Jerusalem to Jericho: A study of situational and dispositional variables in helping behavior" *Journal of Personality and Social Psychology* 27: 100–8.

Dworkin, G. (1988) *The Theory and Practice of Autonomy.* Cambridge University Press.

Elms, A. (1982) "Keeping deception honest: Justifying conditions for Social Scientific Research Stratagems" in Beauchamp, T., Faden, R., Wallace, J., and Walters, L. (eds) *Ethical Issues in Social Science*, pp. 232–45. Johns Hopkins University Press.

Erikson, K. (1967) "A Comment on disguised observation in sociology" *Social Problems*, XIV: 366–73.

Gillespie, R. (1991) *Manufacturing Knowledge: A History of the Hawthorne Experiments.* Cambridge University Press.

Herrera, C. (1999) "Two arguments for covert methods in social research" *British Journal of Sociology* 50(2): 331–43.

Kelman, H. (1967) "Human use of human subjects: the problem of deception in social psychological experiments" *Psychological Bulletin* 67(1): 1–11.

Kimmel, A. (2001) "Ethical trends in marketing and psychological research" *Ethics & Behavior* 11(2): 131–49.

Lawson, E. (2001) "Informational and relational meanings of deception: implications for deception methods in research" *Ethics & Behavior* 11(2): 115–30.

Milgram, S. (1974) *Obedience to Authority.* Harper and Row.

Ortmann, A. and Hertwig, R. (1997) "Is deception acceptable?" *American Psychologist*, 52: 746–7.

Patry, P. (2001) "Informed Consent and Deception in Psychological Research" *Kriterion* 14: 34–8.

Pittinger, D. (2002) "Deception in research: distinctions and solutions from the perspectives of utilitarianism" *Ethics & Behavior* 12(2): 117–42.

Riach, P. and Rich, J. (2004) "Deceptive fields experiments of discrimination: are they ethical?" *Kyklos* 57(3): 457–70.

Saxe, L. (1991) "Lying: thoughts of an applied social psychologist" *American Psychologist*, 46(4): 409–15.

Wolf, S. (1990) *Freedom and Reason.* Oxford University Press.

Index